THE
UNIVERSE
Why Is It
So Complicated?

—

Understanding The Cosmos

by

Larry Eichenauer

Published by
Heritage Publishing.US
Bradenton, Florida

www.heritagepublishingus.com

Contents

Foreword

Why and how was the universe created? Why do humans exist in this vast universe? What is dark matter? What is dark energy? Why are there black holes? Is the universe infinite? Is there other intelligent life in our universe? These are some of the questions that will be discussed in this book.

The greatest minds of our time are often convinced they have resolved most of the questions about the nature of the universe, but then a new technological finding presents a contradiction requiring a restructuring of their conceptual thinking. The more we encounter new discoveries, the more complicated the universe becomes. Physics has become more malleable than any time in history because advanced technology has created more questions than answers. For example, the Big Bang Theory, which was first proposed by Georges Lemaitre almost 100 years ago, often considered to be indisputable, is now being challenged with recent observations by the James Webb telescope.

In the beginning of this book, I look at many of the possible answers for how the universe had an origin from nothing. This may be the most difficult of all questions to answer. Examining the Big Bang and other creation theories will provide a better picture of our universe and help answer some of these most challenging questions.

One of the more intriguing parts of this book will be providing different perspectives for the significance of the human race. What would be the purpose of our universe if there were no humans present to observe and acknowledge its existence? The chapter titled "Cognitive Universe" will provide a different theory not often considered. Could the universe possess cognitive traits? By closely examining the principal components of the universe, this concept could provide new insights for why the universe exists.

One of the greatest mysteries facing physicists has nothing to do with the 5% of visible matter, but instead with the 95% of all the matter and energy in the universe that is not seen or understood. We know it exists, but we don't know what it is. I will explore possibilities to find answers for the mystery concerning the existence of this strange dark matter and dark energy.

The chapter "Black Hole Endoverse Theory" resolves the problem of infinity. This theory suggests how dark matter and dark energy are responsible for creating new universes within a black hole. Only a black hole is capable of duplicating the Big Bang. I will provide the evidence that we are living inside a parent black hole of another universe.

The quantum world is continually evolving as atomic accelerators discover strange new particles. Physicists are still uncertain if the quark could be composed of even smaller particles. Is there possibly no end in sight for how small the quantum world may be? At the other extreme, the universe is continually expanding to a point where there may be no limit to this inflation. Are we living in a dying universe? I will explore the quantum world and make an effort to simplify its complexities. By reviewing all the particles in the atom, it becomes clearer that the quantum world is much stranger than physicists ever imagined.

Gravity is the weakest of the four forces in our universe. Although it may be the weakest, we are far more aware of this force in our daily lives. This weaker force is responsible for our existence. Without gravity, stars and planets would have never coalesced. Physicists believe that this force is produced by hypothetical graviton particles. I will explore how gravitons and dark matter work together to produce gravity. The explanation of how gravity functions may be one of the most innovative parts of this book.

The last chapter examines the significance of the human race. For humans to evolve from stardust would require an extremely complex universe, including our strange quantum world. Dark matter, dark energy, quarks, electrons, boson particles, neutrinos, photons, gravitons, supernovas, black holes, and a planet orbiting an ideal star at just the right position are all required to produce the human race. Without our existence could the universe have been more elementary, and not require all these complex components that make up its structure? Are we the reason why the universe is so complicated?

We are now entering an era of amazing discovery that is beginning to provide some answers in the search for better understanding. What will our civilization be like in one thousand years, or in one million years? Are we alone in the universe? Will we ever meet other intelligent life? I will address these questions and many more.

Chapter I

In the Beginning

"Who are we? We find that we live on an insignificant planet of a humdrum star, lost in a galaxy tucked away in some forgotten corner of a universe, in which there are far more galaxies than people."

Carl Sagan

Every theory that has ever been advocated about the beginning of the universe seems to lose credibility when it encounters the question: "How does something emerge into existence from nothing?" Cosmologists and particle physicists have wrestled with this question for decades. A few physicists have tried to tackle this question by suggesting that the universe was never in a state of nothingness. They propose that when space is completely void of any particles, there is still a potential energy force. But, if this potential energy exists when nothing else does, where did it come from?

"In the beginning".... Everything that exists began with these three words. All we know, see, and imagine had a starting point and that includes God, unless, there was never a beginning and God always existed. Who or what is God? We clearly do not know how to answer this question. First of all, God's existence is based on humanity's respect for the Unknown, and All that has been Created. Many in the scientific community and the Church have asserted that God is the Universe. One could say that the universe was created by an Unknown "Force" that cannot be understood. We refer to this Force as "God".

There are many definitions for God. God is: the Creator, the Architect, the Omnipotent, the Alpha and Omega, the Supreme Being, the Almighty, Lord, Allah, the Father, and the Holy Spirit. All these references are the means by which we, as humans, honor the Creator of our universe.

We do not understand a "Force" that does not have a scientific origin. What is certain, the universe did not begin from anything involving the known forces that are part of the "laws of physics". Only an Unknown Force could be responsible for ALL THAT EXISTS. In physics, there are four forces that control our universe. However, there must be an additional force responsible for the "beginning" of our universe. This force would exist in all dimensions of space and time. Although dark matter and dark energy may be the reason for the development of our universe, this additional force may be the reason for all that exists from the beginning and BEFORE. We can refer to the Ultimate Force as "God" the Creator of dark matter, space, and time.

Many in the field of science base conclusions and theories on proven data. Since there can be no research data that can prove the existence of God, many in the field of physics and cosmology will not surrender to a belief where faith is the prime source for commitment. Instead, many in the scientific community will advocate that the universe was created by chance, or by accident. There are two problems with this conclusion. First, and most important, WE EXIST! When we consider the complexity of the human mind and body it would seem incomprehensible that we would not be part of an Ultimate Plan. We have brains capable of understanding much about the creation of the universe. Our brains have the ability to analyze and solve difficult problems. We have special skills that cannot be challenged by any other biological species.

Our brain contains over 100 billion neurons and over 100 trillion synaptic connections. These allow us to perform unimaginable skills. If human DNA strands could be separated and connected end to end, the strands would stretch to 65 billion miles. Our bodies contain approximately 7×10^{27} atoms. Ninety-nine percent of our bodies are made of basic elements such as oxygen, hydrogen, nitrogen, carbon, calcium, and phosphorus. It is amazing how these 6 basic elements are arranged in a pattern to form the most complex organism on our planet, and possibly the entire universe. If the universe was created by chance, I find it difficult to believe that we were part of this accident. By chance is not a likely conclusion for such an elaborate species called human.

The second problem for denying a Creator is the immenseness, complexity, and beauty of the universe itself. The universe fits together like a puzzle. The web-like pattern formed by the grouping of galaxies, dark matter, dark energy, and even black holes are all part of a cosmic organism that permeates the entire universe. Even the unique composition of the quantum world illustrates how each particle has a purpose for the existence of all things. The complexity of the how atoms and molecules function is far beyond anything we can imagine. The magnitude of formulation that was necessary to create a cosmos this complicated is beyond the consideration of occurring by accident.

In the 1600s, Dr. George Berkeley, an Anglican bishop and philosopher, made the statement: "If a tree falls in the forest and there is no one around to hear it, does it make a sound?" His answer was: "Yes, it will make a sound because God will hear it." What he is suggesting is that the universe would still exist even without the presence of humans or other intelligent life forms. Without humans, who would acknowledge the existence of the universe? One possibility: Could the universe, itself, be self-aware with cognitive ability? Is the universe, God?

When I was a teenager, I read a short story written by Isaac Asimov. The title of the story was "The Last Question". He wrote this in 1956, and he said it was his favorite science fiction story. To summarize, it is a story about a man-made computer called "Multivac". Humanity continually fed data into Multivac, and it ultimately paved the way for humans to travel to different planets throughout the galaxy and beyond. After thousands, and eventually millions of years, humanity became concerned about a dying universe. They kept asking the now extremely advanced "Multivac" computer to find a solution to the increasing entropy that was occurring throughout the universe. Over time, humanity merges with Multivac, and finally Multivac possesses all the knowledge it may need to answer "the last question" posed by humans before they combined with the computer. Their last question was as follows: "How can we decrease entropy before it is too late?" In the far distant future, and as the universe dies, Multivac gathers all the wisdom from eons of information, and just as the last stars begin to flicker out of existence, it believes it has the answer to "The Last Question". Its answer was, "Let there be light!"

How is this short story relevant to the beginning of the universe? It is suggesting that a future technology created the universe. It is saying that for time to have a beginning in an infinite universe, time and space must be created on a future time-line. In this story, the computer was created by humans. The problem here is that both humans and the computer would not exist without there being a Creator of the universe in which humans would exist in the first place. If our Creator chose to create the universe on a future timeline, then this is just another option for the beginning of the universe.

Few scientists will publicly state their religious beliefs. Isaac Asimov was a devout atheist. Some scientists will say they are agnostic. This simply suggests the belief of "maybe or maybe not". Their belief is based on the premise that there is not enough data to support the definite existence of God. And, many scientists who believe in God think of God as a Creator of the universe without any connection to theology. However, most of humanity does not share these views.

I will discuss more on the topics of God, space, and time later in this chapter. First, it is important to understand some of the history of cosmology. Let's review the amazing discoveries over the last century that have propelled science into the modern age of better understanding. Many great astronomers and physicists are responsible for developing some of the most innovative theories of all time. Their research continues to unravel many of the mysteries about the beginning of the universe.

The Big Bang theory has been around for almost 100 years and it still remains the most accepted theory among almost all particle physicists and cosmologists. There are other theories that have spun off of this popular theory. This would include string theory and cyclic theory. Let's look at the most popular theories and then later review how these theories relate to a new theory for creation: "Black Hole Endoverse Theory".

The Big Bang theory was originally proposed in 1931 by Georges Lemaitre, a Belgian cosmologist and priest. At a young age he was interested in both theology and cosmology. In 1923, at the age of 29, he was ordained as a Catholic priest, but his main interest still remained in the field of cosmology. So later that year he enrolled at the University of Cambridge to pursue a degree in physics. After 2 years he decides to relocate to the United States and continue his studies at MIT near Boston. While at MIT he becomes fascinated by the research of Edwin Hubble. During this time, Hubble discovers that the galaxies he has been observing are receding at a progressive rate of speed. Hubble's observations included many objects that were part of the Messier Catalog of unusual stellar formations, nebulae, and galaxies. Charles Messier compiled this catalog from all his observations. Most of the objects he recorded were discovered while continually searching for comets. From childhood he had a fascination for comets. Over a 25 year period, between 1759 and 1784, he cataloged 110 nebulae and galaxies. At the time, he was not aware that many of the objects he recorded were galaxies outside the boundaries of the Milky Way.

It was astronomer, Edwin Hubble, who confirmed the existence of these distant galaxies. Hubble used the newly completed Mt. Palomar observatory to study Cepheid stars in the Large Magellanic Cloud. He discovered that certain Cepheid stars fluctuated in brightness between 2 to 50 days. These stars physically vary in size with a rhythmic pattern. The stars with a longer cycle are hotter stars. By recording each star's duration of change and then comparing the intrinsic brightness to the apparent brightness he was able to determine their distance from earth. When he located these same stars in Andromeda he was able to calculate their distance at approximately 2.5 million light years from earth. This was a major discovery, proving that M31 was a galaxy beyond our own galaxy. He then studied all the Messier objects and discovered more galaxies at even greater distances. By using spectroscopy Hubble was able to obtain more information about many of the stars and galaxies he was observing. When Hubble examined his results he noticed that the more distant the object he was observing, the more the light he received was shifted toward the red portion of the spectrum. With more observations, he concluded that galaxies are moving away at greater speeds the further they were away. In 1929 he made the findings public, stating that the universe appears to be expanding at a progressively greater speed based on distance. This is one of the greatest discoveries during this time.

Georges Lemaitre had suggested 2 years before Hubble's findings that the universe is expanding from the original explosion of a primeval atom. In 1931 Lemaitre published his theory: The universe exploded from a single point and is still expanding at a progressive rate. He said it was "the beginning of the world". Along with Hubble's research, the Hubble-Lemaitre law was established, which stated that galaxies are moving away from Earth at speeds proportional to their distance.

Not all cosmologists supported Lemaitre's theory for creation. One of his biggest critics was Fred Hoyle, a British astronomer, who supported his own "steady state theory". In 1949, while doing a series of lectures on radio, he coined the now famous name for Lemaitre's theory. Hoyle could not accept the idea of the universe exploding from a single point of matter, so he chose to make fun of Lemaitre's theory by calling it "The Big Bang". Hoyle believed that the universe was infinite and he suggested that the Big Bang could not explain how something appeared out of nothing.

The steady state theory simply suggests that matter and energy continue spawning into our universe to maintain their existence at a steady amount, as the universe expands throughout infinity. Old stars and galaxies were continually being replaced by new ones. There was no beginning and no end. This theory, however, lost credibility when the cosmic microwave background was discovered. The CMB was a very significant discovery for those who supported the Big Bang theory.

In 1964, two employees with Bell Laboratories were granted the private use of a horn-shaped antenna that was originally used for linking with the Echo satellites. The Echo satellites were large metallized balloons used for communication, acting as reflectors of microwave signals. After the antenna was no longer needed, Arno Penzias, a physicist, and Robert Wilson, a radio astronomer, were granted the opportunity to use the antenna for their own research. Their goal was to explore the microwave radiation in the Milky Way. They first focused the antenna on a quiet outer edge of our galaxy only for the purpose of calibrating the antenna. Immediately they started receiving the sound of an inexplicable hum. Then, they focused on different parts of the sky and continued getting the same sound. They first thought the antenna was flawed. So, they cleaned all the bird droppings from within the antenna, thinking this was the cause for this unexpected continuous noise. After several checks of their equipment, they came to the conclusion that they may have made the major discovery of the cosmic microwave background. Believing this to be the case, they contacted Robert Dicke, a well known Princeton physicist who was working on locating this CMB at about the same time that Penzias and Wilson were scanning the skies. Even though they were certain about what they had disovered, they needed the expertise of Dicke to confirm their findings. They later confirmed that this radiation was at a temperature of 2.7 degrees Kelvin, just as predicted by Dicke. This established that the CMB was the remnants of radiation left over from the Big Bang, 14 billion years ago.

In 1981 Alan Guth proposed the inflation theory for the first octillionth of a second for the beginning of the Big Bang. He said that hyper-inflation was started by a repulsive force of gravity and negative vacuum pressure, which quickly changed to an attractive gravity force to slow the expansion. All this takes place within the first second.

The Big Bang can be summarized as follows: At 10^{-43} seconds, the four fundamental forces are still combined. The electromagnetic, strong, weak, and gravitational forces are together in singularity. At 10^{-37} seconds the four forces begin to separate. Hyper-inflation begins the expansion. After hyper-inflation, particles and anti-particles collide destroying each other rapidly. By 10^{-6} seconds the particle soup consists of mainly gluons and quarks. Then, the fundamental particles form. Before the first second is completed, massive annihilation occurs between protons and anti-protons as well as electrons and positrons. Quarks are also annihilated by anti-quarks. After a few seconds the temperature drops to 1 billion degrees Kelvin, and when things settle, protons are the predominant particles remaining. At 2 minutes, some funtamental particles begin to combine into deuterium and helium, but many of the protons and neutrons remain uncommitted. Protons remain as hydrogen nuclei. After 380,000 years, these hydrogen nuclei combine with electrons to form hydrogen atoms. The CMB is now apparent, and as it expands over billions of years, it will eventually cool to the present 2.7 degrees Kelvin. At approximately a billion years after the Big Bang, giant clouds of hydrogen and helium atoms begin to coalesce. First generation stars begin to form, and soon after, large groups of stars form galaxies.

The following illustrations provide more details about the Big Bang.

EVOLUTION OF THE UNIVERSE

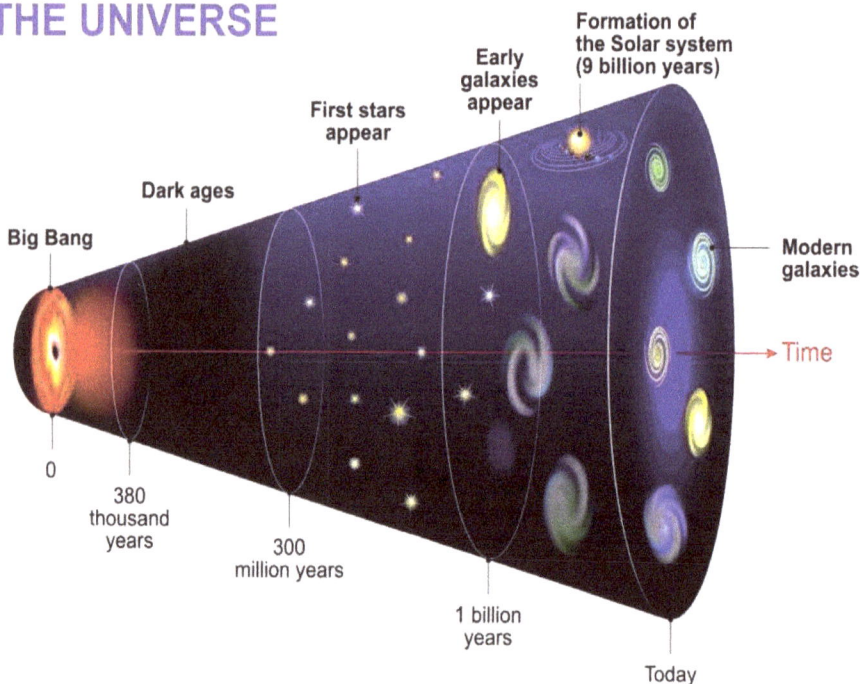

Shown in the illustration above is an era titled "dark ages". This is a period before stars and galaxies began to form. Before the dark ages the universe was a hot soup of particles (mainly protons, neutrons, and electrons). Until atoms of hydrogen and helium started to form, the CMB cannot be detected. This event did not begin until 380,000 years after the Big Bang.

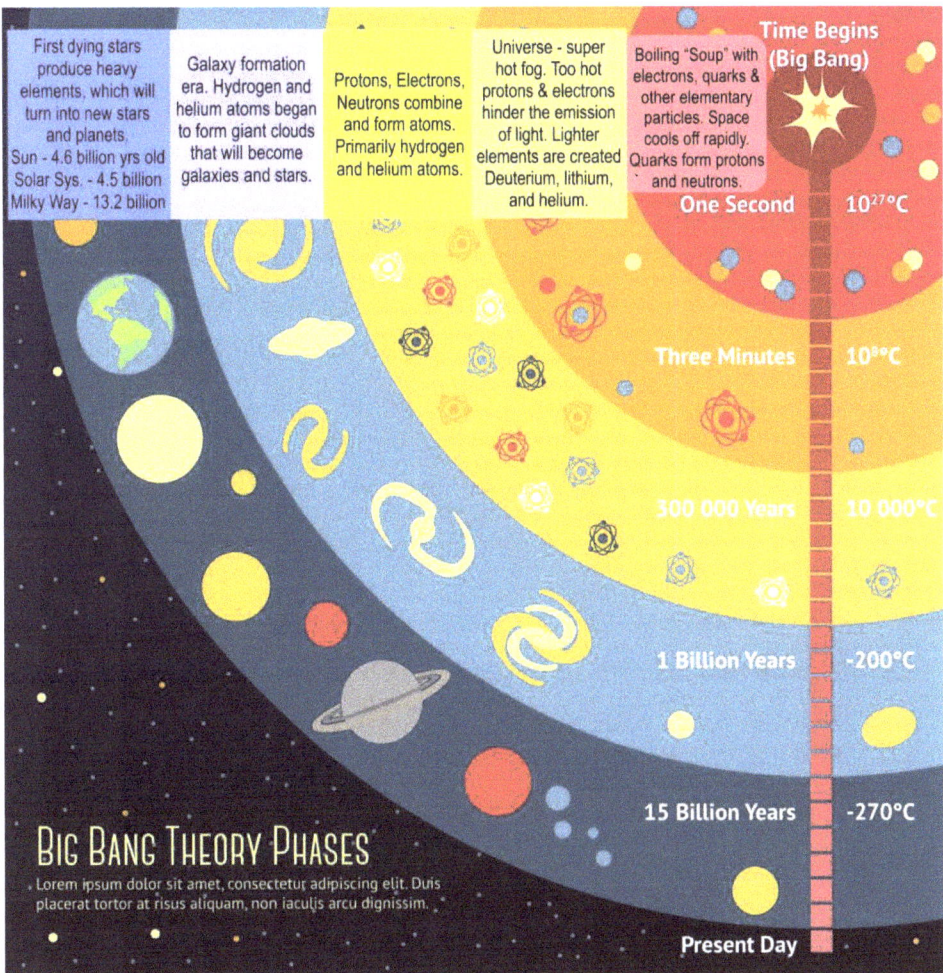

Illustration of the Big Bang Theory

Now, almost 60 years since the great discovery by Arno Penzias and Robert Wilson, the Big Bang has continued to be the preferred theory for explaining the beginning of the universe. There are, however, issues that still need to be resolved before this theory can be accepted as an irrefutable explanation for the beginning of the universe. Although these are opinions, I am basing these conclusions on recent research and discovery.

The following five issues still plague this theory with some degree of uncertainty. First, there is still no solution for how the universe originated from nothing. There have been no valid suggestions for what occurred before 10^{-43} seconds, when the universe was considered in a state of singularity. The only force that has been suggested before all the fundamental forces were combined is that gravity may have existed, or there was an unknown potential energy force that was part of nothing. But, as I said before, this is NOT a definition of nothing. Physicists have called the time of 10^{-43} seconds, the Planck era. Max Planck was a German physicist who is best known for being the creator of Quantum theory. He received the Nobel Prize in physics in 1918 involving wave theory and for his work on atomic structure. So, the first problem is that the Big Bang theory does not include a solution for how something appeared out of NOTHING. What occurred before 10^{-43} seconds?

The second issue is that there has not been any proof for what is called singularity. Space and time totally break down when gravity produces near infinite density. Until physicists can better understand gravity, singularity will only be a "concept" to describe the combination of the four fundamental forces and nothing more.

The third issue concerns the cosmic microwave background. As was discussed earlier, this is the leftover radiation that is the fingerprint for the explosion of the universe. Cosmologists have studied this background radiation thoroughly and for some reason it appears to be too smooth. For stars and galaxies to have developed from the Big Bang era, there should be far more ripples in the CMB composition. Over the last 30 years we have received data from several space probes that have helped in the search for these ripples in radiation structure. COBE (Cosmic Background Explorer) was launched in 1989. Again, in 2001, the US launched WMAP (Wilkinson Microwave Anisotropy Probe) to increase sensitivity for the study of the CMB. Then in 2009, the European Space Agency launched the Planck satellite to provide the most exact information possible for the CMB. So what were the findings of these three celestial probes? First, it was established that the "average" temperature was 2.725 degrees Kelvin. What they discovered, however, was that opposite hemispheres had anomalies in the patterns of temperature variation. They also discovered cold spots in fairly large patches of the sky that counter the standard model. What caused these anomalies is not yet understood. Some have suggested these cold spots may be another universe just outside our own. It does prove that the universe is not the same in all directions, which is different than the standard model. The findings prove that the universe is basically smooth, not much different than the predicted standard model. Another finding by the Planck satellite confirmed the universe is expanding at 67.15 kilometers per second per megaparsec. A parsec is 3.26 light years, so this would be 3.26 million light years. From this information it would mean that the Big Bang occurred 13.82

billion years ago. Planck also determined that the universe consists of 26.8% dark matter and 68.3% dark energy. This means that there is just 4.9% observable matter in our universe, which is slightly more than before the Planck satellite findings. In conclusion, the findings did suggest there were far too few ripples in the CMB structure for the universe to exist in its current state.

The fourth issue is that the universe is expanding faster than the proposed model for the Big Bang, and, there is no end in sight. Some cosmologists have suggested that about 5 billion years ago, dark energy started to force the universe to expand at this greater rate. Until there is a better understanding of how dark energy is the force behind this increased recession, there still remains a question as to why the universe is inflating faster than would be expected.

The fifth issue is based on recent findings acquired by the James Webb telescope. Recent observations have revealed well-formed small galaxies that appear to be from light emitted 400 million years after the Big Bang. This conflicts with the standard model, which suggests that galaxies could not have formed this soon after the Big Bang. According to the standard model, galaxies should not form until at least one billion years after the Big Bang. This again is a question that science has no answer for, since the findings from the Planck satellite reveal the age of the universe to be 13.82 billion years. If the Webb telescope is able to look back even further than 13.4 billion years and locate more of these well-formed galaxies, some major questions will need to be addressed.

In the third chapter I will be proposing a different theory that will focus on these five issues. The next theory for creation I would like to analyze is the cyclic universe. This theory was proposed some 2000 years ago by Hindu philosophers. Basically this theory takes the Big Bang theory and adds an "infinity clause". Instead of the universe expanding forever, the universe somehow acquires a special energy or gravitational force which would stop the expansion, and as a result, it eventually starts to contract. After billions and billions of years it contracts to a near infinite point, which is called "The Big Crunch". The universe then rebounds to repeat the entire process of another evolving universe. This cyclic pattern would continue for eternity.

Well-known theoretical physicists, Paul Steinhardt and Neil Turok have refined this theory to help fill in the gaps that prevented this theory from originally gaining a foothold among cosmologists. They suggest that dark energy will eventually change the nature of the expanding universe. This energy will at some point stop the expansion and cause the universe to begin contracting. Once it contracts to a near infinite point of matter it will encounter singularity and the Big Crunch occurs. They are suggesting a slightly different approach for achieving singularity by proposing that a membrane in our four dimensional universe collides with other membranes existing in other additional dimensions. Each one of these membranes possesses the characteristics of matter in each, and once this collision occurs there is a transfer of matter and energy to produce the Big Crunch.

This does not explain how the membranes came into being. It does create another possible consideration for the beginning of the universe. The concept of membranes was originally proposed in the Superstring Theory, which includes 10 dimensions instead of four (3 dimensions plus time).

What issues need to be resolved for cyclic theory? First, until we know how dark energy affects the expansion of the universe, it will be only conjecture to consider that this unknown force will somehow stop and reverse the current trend. Secondly, it is difficult to imagine the chaos of a contracting universe. When the universe inflates, everything is evolving and the space for this evolution is virtually unlimited. But, when you reverse the process, things really get wild as stars, galaxies, and black holes begin to collide. Black holes could create a totally different type of universe for future cycles by consuming most of the universe before it reached a critical diminished state. Rewinding a universe is not just pressing a button and everything simply reverses identical to the forward motion. This reversal would not end with any predictable success. The chaos could totally destroy the likelihood of a Big Crunch.

Thirdly, the same issue exists for this theory and all previous theories. How did membranes, other universes, and multi-dimensions exist before the Crunch? How does something originate from nothing? The Crunch theory is infinite into the future, but not into the past.

All these theories provide some answers for the beginning of the universe, but there are still many questions that need to be resolved before the cosmic puzzle begins to make sense. Once we have a clear understanding of Dark energy and Dark matter, all the pieces to the puzzle will begin to fit together. We must find the hidden solutions to these mysteries. Dark matter consists of 85% of the matter in the universe, and we can't see it. And, we have not been able to discover its existence in the quantum world. How is this possible? We know it is there by how it affects the light we receive from distant galaxies. We know it was responsible for the formation of stars and galaxies. So why is there such mystery behind its existence? Dark matter must permeate the entire fabric of space. When considering the significance for its existence, dark matter is the "lifeblood" of the universe. I will discuss this topic in great detail in the chapters, "Dark Matter and Dark Energy" and "Quantum World".

Carl Sagan had a memorable phrase: "We are all made of star-stuff." The universe made life possible on this remote planet called Earth. Everything in the universe is interconnected and everything has a purpose. Stars explode, creating gas and dust for new stars and planets to form, no different than our solar system that coalesced approximately 4.5 billion years ago. All the elements necessary for life to evolve on our planet originated from exploding second and third generation stars. Even monster black holes have a purpose other than swallowing stars. Most of the larger galaxies formed with the assistance of the gravity from black holes. It is possible that black holes may provide the solution for how the universe began. Until science has a better understanding of how everything in our universe functions, the beginning of the universe will continue to remain a mystery.

At the beginning of this chapter I addressed the question of how the universe evolved from nothing. The only way to avoid this so-called "dead end" philosophy is to accept the reasoning that infinity is the best solution to the problem. It is impossible for the human mind to comprehend the meaning of infinity. Everything we can relate to in the universe is born and eventually dies. However, when we consider what we don't see, everything is different. According to the findings from the LHC (Large Hadron Collider), protons and electrons have been observed to not decay. Although, there have been a few physicists who have suggested that these fundamental particles may decay over trillions and trillions of years. If we go with current data, these fundamental particles live forever (infinity).

We often hear it said. "God is infinite." Or, "The universe is infinite." What are we saying? In reality, we are simply admitting, we do not know, so we therefore make the assumption – infinity. If infinity exists for space and matter, it also must exist for "time". When discussing the universe, time is synonymous with infinity. One cannot help but imagine the scope of such a vast universe without thinking of time being infinite.

The title of this chapter is "In the Beginning". But, there may be a problem in the use of these 3 words. If the Universe is infinite, all matter and energy have always existed. There was never a time when nothing existed. We may say that God is the Creator of the Universe, but instead it has always existed, which might suggest that there was never a beginning. How is this possible? Science suggests that the Big Bang is the moment of creation, but where did all matter and energy come from? Everything in the Universe did not just suddenly appear out of nothing! Even if there are an infinite number of universes, how can we say ALL that exists just popped-up out of nowhere? This is why the Universe must be infinite. So, if we accept infinity to mean that the Universe has always existed, we must view God differently. God IS the Universe! One thing that must be considered about creation: It is continuous and infinite. New universes are being created, and this is how God is the Creator and Sustainer of the Universe.

To summarize this reasoning for an infinite Universe, it can be stated that the entire Universe includes an infinite number of universes that are continuously being created to preserve its infinite existence. The Big Bang is how infinite matter and energy are transformed to continue this cycle.

At 13.8 billion years old our own universe is fairly young, and its existence appears to be finite. The limited existence of our universe poses another question. Can infinity and finite coexist? We can suggest that many things may be infinite, but some are not, because of the laws of physics. Since the laws of physics exist in our universe, we can say that infinity does not apply to everything in the Universe.

For a better explanation of infinity, it is best to think of time on a time-circle, instead of a timeline. When a point is plotted on the time-circle, a moment in time is created. A point positioned on the circle determines when "now" exists. By applying the laws of physics, the movement of time will always move toward a future event. Time does not require an observer for its existence. Time, space, and matter have always existed on the time-circle, which means they are infinite. A past or future frame of reference cannot exist until there is a point of origin (now). The concept of nothing cannot exist on a time-circle. If time is infinite, "now" has always existed.

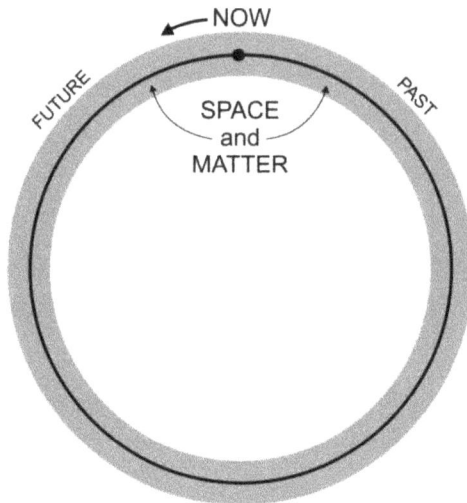

If the Universe is God and the Universe is infinite, space, time, energy, and matter have always existed – There can be no beginning in an infinite Universe.

Throughout the Universe, all that exists may have originated and then developed by evolution. The complexity of the Universe may have evolved through trillion of years of change. Even our own universe is continuously changing. The early Universe may have been far different from what we now observe. The Universe has perfected its structure and functions to become so complicated that it may never be fully understood. The universe we know may be beyond human comprehension, but the Universe we don't know may be "infinitely" complex. The Entire Universe may include an infinite number of universes, including unlimited dimensions of space and time. Our search to understand the "The Theory of Everything"* may be futile, but at least we will enjoy the pursuit.

Whether there was a beginning of the Universe or it always existed doesn't really matter. Because of these complexities, the existence of the human race is part of this enormous Universe. We should feel special knowing that all that we see and experience has evolved over trillions of years, providing humanity the chance to be a part of this unfathomable Creation.

*References book by Stephen Hawking

Chapter 2
Cognitive Universe

"Time is nature's way of keeping everything from happening at once."

John Archibald Wheeler

Without humans to acknowledge its existence, what would be the reason for the universe to have been created? As humans, we believe that there must be a purpose for everything we do and everything we visualize around us. Why would any Power or Force create a complex universe that had no purpose for its existence? A universe "this complicated" cannot be the result of a mistake or accident.

The universe we know has existed for at least 13.8 billion years. The presence of humans or any intelligent life form would not have existed for at least the first 7 to 8 billion years. So if there is a purpose for the existence of the universe it may not have included humans in the original Plan. Therefore, this would suggest that, for the universe to have a reason for its existence, there has to be another way for the universe to be acknowledged. Could the universe itself have the ability to be self-aware?

To determine if our universe possesses cognitive traits we need to look at the evidence to support this theory. The following topics will be addressed to confirm whether or not the universe has the ability to be self-aware: cosmic web, dark matter, cosmic microwave background, dark energy, interconnections, and self-replication.

The most obvious support for the Cognitive Universe Theory is that everything in the universe has a reason for existence. It is ALL interconnected. What are some of these connections? First, the explosions of large stars provide the seeding of heavy elements that are essential for creating solar systems and life. Our own solar system exists as a result of likely three generations of stars that exploded, releasing heavy elements from their cores. Approximately 4.5 billion years ago, our Sun and planets were the by-products of these early stars. This didn't happen by chance. The universe has a plan with a primary focus on evolution. The universe, including everything on planet Earth, is constantly evolving to secure its existence. Just think about the evolution of our planet in order that it can support life. Its atmosphere and magnetic field provide the envelope of protection from harmful radiation from our Sun. The atmosphere contains just the right amount of water and specific gases to maintain an ideal climate for life to flourish. Throughout our universe powerful supernovas may harm nearby planets, but at the same time this violent act may stir the pot of nearby nebulae to help in the formation of new stars and planets. In this complex universe there is ongoing cause and effect. Again, it is ALL interconnected.

The pattern of the cosmic web is mesmerizing. When one considers the intricate shapes and the woven design of galaxies and filaments of gas and dust, you have to contemplate the reason for its existence. These unusual formations must be part of a plan to maintain a relationship between the sum of all the parts of the universe. This abstract pattern can be compared to neurons, dendrites, and axons in our brain. There may be a reason for this apparent similarity. What is the purpose of this complex pattern of cosmic material? Dark matter appears to be responsible for stabilizing the positions of the galaxies and surrounding matter. Dark matter is the "glue" that is responsible for this unique continuous meandering design. Is it possible that dark matter created an intricate cosmic web to provide a connecting network between the galaxies? This cognitive theory may seem unrealistic, but it is possible the universe has a greater significance other than for only humans to acknowledge its existence. If we refer to God as the Universe, this connecting network would suggest a cognitive parallel. **The following illustration** represents how all the galaxies and groups of galaxies are connected by a common thread, the cosmic web.

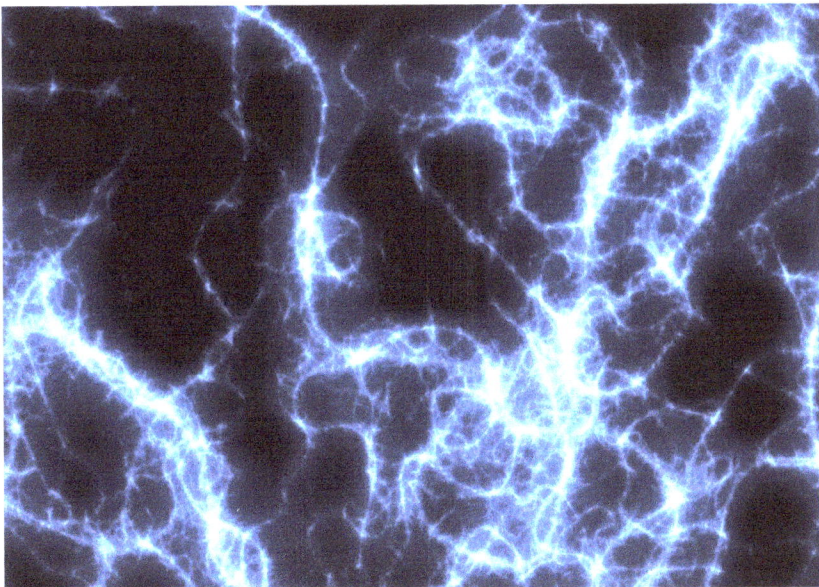

ILLUSTRATION OF THE COSMIC WEB

If dark matter is the "glue" that holds the galaxies and the universe together, what is the reason for the existence of dark energy, which is the force that is pulling the universe apart? Dark energy is acting like a cancer that is causing the universe to expand at an ever increasing rate. Why does this force exist if it is destroying the structure of our universe? Dark energy is pulling on the filaments that hold the cosmic web intact. Up until now I have suggested that violent forces, such as black holes and exploding stars, have a purpose by promoting the formation of new stars and planets. Black holes are also responsible for forming and maintaining the structure of galaxies. Dark energy, on the other hand, does not appear to have a positive role in our universe. However, dark energy may have been the driving force for the birth of our universe. This has yet to be proven, but considering it is a major part of 95% of the universe we know very little about, it had to have played a major role from the very beginning.

Einstein's theory of general relativity explains how spacetime is curved by matter producing a force of gravity. But, what happens if space itself is stretched by unknown forces? Dark energy appears to be the reason for the universe to be expanding at an ever increasing rate. We call it dark energy because it appears to be the unknown unstoppable force propelling this expansion. Maybe we are not considering the more obvious cause for this expansion. Dark energy may be space itself. We may be using the incorrect terminology for dark energy. The real cause for this expansion may be that space is being stretched by an outside force, or expansion is just a natural force common to space itself. If the Big Bang started this inflation, then you would expect it to continue, unless a greater force can reverse the process. Dark matter may be the opposing force, but it obviously cannot stop the current expansion of space.

Until about 5 billion years ago, the universe was expanding at a slower rate. For some unknown reason this expansion started to increase at an alarming rate. We are not sure what caused this change, but, it is apparent in the cosmic microwave background information, provided by the Planck satellite. So what might be causing this aggressive expansion? There can be only two possible answers. Either the force of dark energy is getting stronger, or there has to be outside forces affecting our universe. I mentioned previously that dark energy could be space itself. The expansion of space could be a natural characteristic for our universe. But this does not answer the question for the sudden increase in expansion almost 5 billion years ago. This would suggest an outside force may be responsible for this greater expansion rate.

The first possibility is that the void of space that the universe is expanding toward acts like an attractive force. It would be like air leaking into a vacuum. Space is being pulled by a "growing vacuum" in the void of space. Secondly, think of space being stretched over a greater area. The lines of space would continue to separate by spreading further and further apart. This separation would also affect Einstein's theory of general relativity by reducing the impression of gravity in the contour of space. If stretched space reduces gravity then expansion will increase at a rate proportional to the distance. The stretching of space will continue to reduce the influence of gravity. Mass does not change, but by stretching space it will have a reduced gravitational effect, proportional to the amount that space is stretched. The combination of greater vacuum and the reduced gravity will cause space to expand at a greater rate.

There is also one other possibility for this expansion. If our universe is surrounded by other nearby universes, it could eventually approach one of these nearby universes, causing space to be stretched by outside gravitational forces. This close encounter could greatly alter the structure of our universe. Quite often galaxies in our universe collide. Even our own Milky Way galaxy will collide with Andromeda in approximately 4 billion years. Most likely our universe is not colliding with another universe because the effects would be obvious. The collision of galaxies creates enormous chaos in their stellar structures. If two universes collided, the chaos would be monumental. Most likely our expansion would be the result of a close encounter. However, this does not rule out that the close encounter might later become a collision of two universes.

Is there a connection between dark energy and the theory I have been presenting? If dark energy is space, then it plays a major role in this theory. Other than time, it is the most important part of the universe. It is the membrane that supports and holds all matter in a specific place. Without it nothing could exist. Dark energy also plays a major role in black holes. However, this is where it all changes! I will explain this interconnection at the end of this chapter.

Our universe appears to be expanding to the point that it may eventually tear apart the threads that hold the cosmic web together. Trillions of years from now, galaxies could become lonely islands separated by countless light years, where-upon, the light that once lit up a bright sky will become dark and empty.

We want to believe that our universe will go on for all eternity. But it is now becoming more apparent this is not the future of our universe. How does a dying universe fit into any plan where everything has a purpose? Obviously, this is not what common sense tells us. But we all know that in our own lives, things occur that just don't make sense. Why do destructive events occur and create unwanted consequences? However, quite often destructive issues may transform into positive results. While it may not be obvious at the time, a negative occurrence may generate a turn of events that produces a positive outcome. Dark energy may follow this same unpredictable pattern.

Dark energy may be responsible for destruction, but its ultimate positive purpose may be creation. For our universe to be aware and possess cognitive traits, it would need to replicate its own existence to prevent its eventual demise. If we search the universe for how this replication could take place, there is only one possible answer, a massive black hole. We know they exist at the middle of all the large galaxies in our universe. They appear to be swallowing surrounding dust and gas and sometimes entire stars. Because our planet is nearly 26,000 light years from the center of our galaxy it is difficult for us to examine our own massive black hole, Sagittarius A*. This black hole has the mass of 4.5 million suns, which is small compared to some black holes that have masses equal to billions of our Sun. These are called supermassive black holes and are visible from billions of light years away. The energy from these huge black holes is so violent that they often shoot large plumes of star matter into space, making them visible from unimaginable distances.

In order to prove that dark energy is responsible for how our universe can spawn new universes, it is necessary to understand more about black holes. Albert Einstein suggested in his general theory of relativity that something on the magnitude of a black hole should exist. In 1916, Karl Schwarzschild proposed the idea of black holes using some of Einstein's relativity equations. It was physicist John Archibald Wheeler who first used the name "black hole" in 1968. He suggested that large stars could collapse by the force of gravity at the end of their life cycle and become a black hole, where even light could not escape. Black holes were only an equation until 1971, when British astronomers Louise Webster and Paul Murdin at the Greenwich Observatory observed an invisible massive object circling a large blue star about 6000 light years away. They called this object Cygnus X-1 and determined this invisible object to be a black hole. It wasn't until 2020 that three great physicists, Roger Penrose, along with Reinhard Genzel and Andrea Ghez, received the Nobel Prize for their discoveries and studies of black holes, more specifically, the massive black hole at the center of our Milky Way galaxy. Considering this Nobel Prize was presented only recently, it is apparent we are still in the process of learning how these monster black holes function. Other than dark energy and dark matter, we have more to learn about black holes than anything in our universe.

We must decipher what we now know about black holes before tackling the cognitive theory in more detail. Black holes are mysterious. They possess a protective horizon layer. Nothing passes through this threshold unless it approaches the unknown moment of singularity. I use this term of singularity loosely because we really have no idea what happens in the quantum world at that point. Black holes are different than heavy neutron stars. Neutron stars do not have the protective outer layer we call the horizon. The way the horizon functions is that it appears to create a barrier to the inside of the black hole. By synthesizing the atoms in this protective layer it opens up many possibilities for the inner core of the black hole. The force of gravity normally has little or no affect on quantum particles, but when it reaches a threshold wherein gravity competes with the strong force in the nucleus of the atom, all types of possibilities must be considered.

Stephen Hawking suggested that when "stellar" black holes are not continuously accumulating new matter from their surrounding space they will expel radiation (Hawking Radiation) from their horizon layer. Over time this type of black hole will release all of its energy and mass, eventually evaporating into nothingness. This is not yet proven to be the ultimate fate of stellar black holes. There is the possibility some dying matter would still remain where the stellar black hole once thrived.

Massive black holes follow a different scenario, and because they live at the center of galaxies they possess an endless food supply. It continues to grow larger around its horizon as it gobbles up surrounding stars, dust, and gas. Some supermassive black holes have the mass of over a billion suns. The largest known supermassive black hole, TON 618, has a mass of 66 billion suns. Fortunately, it is located 10 billion light years from Earth. This spinning black hole emits astonishing flashes of gamma rays and light producing what we call a quasar. Its luminosity is equivalent to 140 trillion suns. The black hole at the center of the Milky Way is small in comparison, having a mass of 4.5 million suns. These massive black holes are the reason we have galaxies in our universe. About 250 million years after the beginning of our universe, black holes began to form by gravity and dark matter pulling together large masses of hydrogen and helium gases. They developed from much larger areas of gas than stars. Thus, as they formed, the gravity was so great that they became black holes instead of stars. Some black holes evolved from young super large exploding stars, whose cores shrunk quickly to become medium sized black holes. This occurred after the massive black holes coalesced. Many of the medium black holes collided with, or were gobbled up by the supermassive black holes to produce bigger black holes. All these massive black holes possessed enough gravity to collectively assemble galaxies like our own Milky Way. All this evolutionary process did not happen by chance. The universe includes an organized Plan that exceeds our imagination. The following artistic renderings of black holes exemplify the enormous power they possess. What lies within the horizon of these massive black holes may hold the secret to the beginning of the universe.

MASSIVE BLACK HOLE

BLACK HOLE QUASAR

The biggest unknown is what we can't see. What happens to matter, space and time after they cross through the black hole event horizon? What is the structure, if any, on the inside of a black hole? First, we know black holes are very dense. To achieve density similar to that of a black hole, it would require shrinking the entire Earth down to the size of a marble. Just try to imagine this for a second, all of the Earth inside of a marble! So how big are these black holes? A stellar black hole that has a mass of 20 times our Sun would be 10 miles in diameter. The massive black hole at the center of our galaxy is approximately 15 million miles in diameter. Our Sun is slightly less than 1 million miles in diameter. The diameter of our own black hole is equivalent to the width of 17 Suns.

Let's return to the question of what is inside the horizon of a massive black hole. As I mentioned earlier, black holes are different than anything else that exists in the universe. Neutron stars are the next closest density. This would be like shrinking the earth to the size of a football field. A neutron star consists of almost 100% neutrons. The pressure inside the star is so great that the electrons and protons are squeezed together causing the negative charged electrons to combine with the protons to produce a star that consists of almost entirely neutrons. The space between the neutrons is much smaller and only the strong force prevents the star from becoming a black hole. Theoretical physicist Aleksi Vuorinen has suggested that the innermost core of neutron stars may consist of a soup of quarks and gluons.

What happens inside a black hole is still a mystery. Once matter crosses the threshold of the event horizon, the laws of physics break down. Many physicists suggest that singularity is the only possibility, wherein all four forces combine into a superforce. What is a superforce? Some consider it pure energy. If the laws of physics do not apply to the core of a black hole, all that is left is imagination.

To answer a complicated question, it is best to consider only what appears obvious based on basic principles. If certain activities follow a trend, then guidelines are formed to provide a predictable result. In other words, if evolution follows a specific sequence of events, then a similar evolutionary pattern should occur when circumstances are similar in nature. If stars and galaxies formed by coalescing gas and dust, then we can theorize that stars and orbiting planets must have formed in a similar fashion. This may have required the assistance of dark matter. Throughout the entire universe, evolution plays a common role. This is the most prevalent phenomenon in the universe. With this premise, it is necessary to conclude that black holes are not just a freak of nature. Their mystique strongly suggests they must play a significant role as part of a bigger Plan.

If evolution is the dominate role of the universe, we can only conclude that black holes must play a part in the continuation of this principle. Recently, several cosmologists have suggested multiverses are created within black holes. In the past this would have been only science fiction. But now, with all our technology, we are able to answer some of the most difficult questions about the universe. Physics is being pushed to the limit to try to resolve some of the mysteries that seem to not match the standard model. In the past, only a few theoretical physicists supported some of these unproven concepts. Now, that number has grown because mathematics cannot supply answers when the laws of physics are being challenged. In addition to the multiverse theory, there are a few cosmologists who support a theory that within a black hole there is an entire universe on a scale that is similar to our own universe. For some this may appear to go well beyond the laws of physics, but there are many good reasons to support this theory.

First of all, this theory will not work based on our current understanding of physics. If the universe is to exist for all eternity, there has to be a plan to accomplish this. Where else can you imagine a better plan for "self" preservation than a black hole? The black hole is the only object in the universe that has the ability to replicate the "Big Bang". Let's look at what would be necessary for the inside of a black hole to evolve into a new universe. One element that has to be part of this equation is dark energy. This unknown component may be the secret for how this theory can become reality. Dark energy may be a force, but its true nature is really just space/time. So, what has to happen for matter to cross the event horizon and enter the depths of the black hole? To create a better description of the process it is important to change the name for dark energy, which is not the best title for this form of space/time. The new name is "SPACE ENERGY". What is inherent to space energy is that, it has the ability to either expand or contract. In our universe we are seeing the affects for how this space energy can cause the universe to expand. I discussed this earlier in the chapter. Now it is time to look at how all space and matter within the horizon of a massive black hole can be shrunk to the quantum level.

Space energy has the potential to reduce space and everything in that space to a near infinitely small quantum structure. This is suggesting that our entire universe would exist inside a massive black hole. We are living inside a massive black hole in a parent universe that was created 14.8 billion years ago, including one billion years for the black hole to evolve within a galaxy that may be similar to our Milky Way galaxy.

The horizon of the black hole is the protective layer that has the potential to separate dimensions of space. So, what can happen along the event horizon that could change normal space to quantum space? The horizon of the black hole is the area where all of matter is transformed into a quantum soup of unknown particles. This is directly related to energy, density, and temperature. This transformation for flexibility will only take place when temperatures are near one quadrillion degrees Kelvin. Those particles of matter that are not transformed remain closer to the outer edges of the horizon where temperatures are slightly less but still extremely hot. This outer layer is pressurized by the hotter temperatures below them, causing the matter to explode into space along the axis of the spin of the black hole. The lower level consists of extremely hot malleable particles that are vulnerable to gravity and space energy. Space energy is eventually overcome by the force of extreme gravity. At this point, space starts to compress to a near infinitely smaller structure, while the space energy pulls with it the hot malleable soup of potentially infinity smaller particles. Once space energy has regained its equilibrium with gravity, the compression of space and soup of particles cease this "free fall" event. When the space energy slightly overcomes the force of gravity, there is a recoiling effect (Big Bang) and as the temperature cools within the first second, the particles begin to reform, but now in a new quantum state. This is the "Big Bang" of a new baby universe inside a black hole. In the first chapter I discussed the Big Bang in great detail. The Big Bang evolution would still remain the same, but now it occurs inside a black hole.

The next chapter will provide the details to validate the Black Hole Endoverse Theory. This theory does provide a means for the Big Bang to counter the expansion problem, and thus, the universe will remain infinite. With this theory, the number of dimensions that exist in the "entire" Universe would be endless. A "smart" universe must find a way to replicate itself if it chooses to survive the potential of death and destruction. Dark energy may appear to be the villain, but in the Ultimate Plan, it may possess the greatest power of all. It may be the reason for the beginning of our universe. Along with dark matter, it also helps maintain balance and structure for "most" of the lifespan of each universe.

Perceiving the Big Bang within a massive black hole may be difficult to imagine. However, it is an ideal solution for maintaining an infinite existence. One must consider, is there any other region in our universe wherein the Big Bang could occur? The only answer based on our current understanding is NO!

All that has been presented in this chapter supports a complex Ultimate Plan that would suggest the universe must possess cognitive abilities. We discussed how the cosmic web is an intricate connection between the galaxies. We looked at how dark matter is mainly responsible for the formation of the cosmic web, galaxies, stars, planets, and black holes. I also illustrated how everything in our universe is interconnected. We looked at how the universe is the result of constant evolution, a trait necessary for infinite survival. And finally we looked at the formative characteristics of "space energy" and how it is responsible for the creation of new universes inside black holes. These discussions should provide all the evidence needed to confirm the cognitive power of the universe. The universe does not think like humans, but it has the ability to create energy, as well as form and maintain structure. The function of the universe is only possible because of how the particles in the quantum world work together with the unknown forces of dark energy and dark matter. Every portion of our complicated universe has a reason for existence. Complexity is never the consequence of chance!

Chapter 3

Black Hole
Endoverse Theory

"Black holes are where God divided by zero."

Albert Einstein

Throughout this book it is the goal to consider all the possibilities, but always attempt to maintain the laws of physics. However, these laws do not always provide the best answer for a problem. There are several issues that need to be addressed because physicists are being challenged by forms of matter and energy that don't fit into the standard model, such as dark energy, dark matter, neutrinos, black holes, aggressive expansion, singularity, gravitons, and quantum entanglement. This list does not include all the uncertainties we face in physics, but it validates, we have much to learn. With recent technology we are beginning to slowly develop possible solutions to some of these mysteries, but it could be centuries from now before we can firmly say, "we have it all figured out". There is the possibility that some things will <u>never</u> be resolved if the solution exists in another dimension of time and space. Many problems are solved by considering the most practical and obvious solution. The answer has been right in front of us all along. We just need to put the pieces of the puzzle together in the correct order.

I chose to start this chapter with this introduction because we may not be looking at all the signs along the highway to reach our destination of understanding. The universe appears to have been created wherein everything is interconnected and has a purpose. Everything seems to fit into a mold where the best answer is the most logical possibility. The most mysterious component of the universe is the existence of massive black holes. There is something really different about the structure of a black hole that is unlike anything else in the universe. You have to ask the question. "Why do we have black holes in our universe?" If their purpose is to improve the way gases can coalesce to form stars and galaxies, it would make more sense to let dark matter be the ideal force. A strange and complex black hole would not be the first choice for creating a galaxy. If you follow this course of logic, there has to be a specific purpose why black holes exist and it does not include just galaxy formation.

One of the most obvious and unusual characteristics of a black hole is how matter and energy accumulate at a horizon, rather than like many stars, which after exhausting all their "fuel", collapse and explode as supernovas. The answer to this black hole mystique lies in the apparent horizon and event horizon of its outer shell. If we can better understand what we can see, then we can formulate our best predictions for what we can't see.

In the previous chapter I presented the concept of how a black hole could be the embryonic portal to the evolution of new universes. If certain black holes are the incubators for new universes, it then suggests that our universe is currently growing inside a black hole of a host universe. If we begin to compare the black holes in our universe with our visible universe, we will discover enough similarities to validate a strong probability for the acceptance of this theory. In order for this theory to be explained, I will often make comparisons to our own universe.

There are two types of black holes, stellar black holes and massive black holes. Massive black holes are located at the center of almost all galaxies. There are a few massive black holes that have been found outside the center of a galaxy but it is believed these black holes were pulled from their center by a near collision with another massive black hole. There are likely 10 million "stellar" black holes in our own galaxy. They are the end result of large collapsing stars. Some stars have a lesser mass and become white dwarf stars when they use all of their fuel to support their outer layers. The Sun is an average star that will one day use all of its hydrogen fuel at its core, after which it will start to burn hydrogen at the higher levels, causing it to expand into a red giant. The Sun will engulf the Earth about 5 billion years from now. When most of the hydrogen fuel is gone, it will collapse, releasing its outer layers into space, giving the appearance of a ring nebula surrounding our smaller Sun. After burning all of its hydrogen it will ultimately condense into a white dwarf.

Stars that contain 3 to 10 solar masses will often die a different death. These stars burn through their fuel much faster, and once they burn up all their hydrogen they will collapse and explode as supernovas. These stars will become neutron stars or black holes. If the stars are at least 10 solar masses these are the stars that will likely become stellar black holes.

Stellar black holes are located throughout our galaxy and are often difficult to locate. A stellar black hole with significant mass could gobble up our solar system, or at least throw the orbits of the planets into total disarray. The only way to see black holes is by observing how their gravity affects surrounding stars, or by visually seeing them devour nearby stars. If a black hole 30 times the mass of our Sun would pass within the distance of Neptune, it would be the end of civilization. Fortunately the nearest black hole we have been able to locate is over 1500 light years away.

Stellar black holes are different from massive black holes at the center of galaxies. Stellar black holes are a denser version of neutron stars. Some neutron stars can become a black hole by accreting additional mass from nebulous surrounding gas, or by colliding with another neutron star. Another characteristic of a black hole is that they spin at incredible rates. For example, the black hole at the center of our galaxy rotates on its axis over 1150 times per second. This is near 50% the speed of light. It is difficult to imagine something with the mass of 4.5 million suns and a diameter of 14.6 million miles spinning 1150 times every second.

How are stellar black holes and massive galaxy black holes different? The biggest difference is mass obviously. There is a huge gap between the mass of stellar black holes and galaxy black holes. Large stellar black holes may be 3 to 50 solar masses, whereas galaxy black holes are 3 million to 66 billion solar masses. The formation of galaxy black holes may date back to when galaxies coalesced. Most massive black holes began to develop within the first 250 million years after the Big Bang. This is likely the greatest difference and may suggest that these two types of black holes may share fewer similarities than expected. Galaxy black holes likely formed by extensive accretion, or by collisions with other massive black holes. Since the formation is so different, massive black holes should be given a different name. The structure of the massive black holes may be more different than predicted. The massive galaxy black holes also acquire huge accretion disks around their center. Large stellar black holes may generate an accretion disk but it is almost insignificant due to its much smaller mass since these black holes have little to feed on. The black holes at the center of galaxies have nearly an endless supply of matter to devour. Both types of black holes possess an event horizon, but this may be the point where things start to differ greatly.

If we consider these comparisons, what stands out is the process of formation and the amount of matter they ingest. Basically, stellar black holes are dying stars with little purpose and interaction. Galaxy black holes appear to have several purposes for existence, and have extreme interaction. Galaxy black holes are so unique that their structure must exhibit extraordinary benefits. Some of the observed reasons for their existence include: gravity assistance for galaxy formation, emission of elements from its accretion disk, and radiation to promote the birth of new stars. The last and most important reason for existence is replication, creating a new baby universe.

Considering that our expanding universe is not going to live forever, there must be a means to preserve its existence. We have discussed different theories and none provide a solution to this problem. The Big Crunch is being discredited by how dark energy is causing our universe to expand to oblivion at an ever increasing rate. The steady state theory does not provide a solution for the replacement of matter in the vacuum of space as it expands, and it does not offer an origin for the universe which contrasts with the cosmic microwave background map. The Big Bang has now become the only accepted theory, but the aggressive expansion is not part of the standard model. If we accept most of what the Big Bang has suggested, we are still looking for a solution to the preservation of the universe. When we observe all the likely possibilities for a way to preserve this existence there are very few remedies. Only a massive black hole has the ability to duplicate what occurred 13.8 billion years ago. Our "Big Bang" emanated within the depths of a black hole. Is this possible? It's time to take a journey into a black hole to find the answer to this question.

The best way to start this journey is to first imagine that we are currently living inside a massive black hole. Since we know black holes are extremely dense and should be filled with distorted matter that has passed through the event horizon, how could our universe appear the way it does to us with all the space between everything we see in space? First of all, we do not understand singularity. At singularity all matter and energy exist in a different form. This new quantum world is not the world we see in our labs or at CERN. Singularity consists of matter that has been reduced in structure by a factor of 10^{45}. If everything including quantum particles were shrunk by 10^{45} power, our entire universe would be the size of a quark inside a black hole. However, you may wonder, where is all the matter that is constantly being consumed by the black hole? Only as the black hole was first being formed, is when the Big Bang occurred inside a parent black hole within a 10^{45} larger parent universe. This is the time when our universe was born. Singularity only happens once and for a split second early in the formation of the black hole in our parent universe (approximately 1 billion years after our parent black hole formed). Our universe is 13.8 billion years old and our parent universe is approximately 13.8, plus 1 billion years, which equals 14.8 billion years. This age would be on our level of observation, but time was also altered when our universe was created. Our parent universe could be much older in another dimension of space and time.

How does our expanding universe compare with what is happening in our parent black hole? Is there a relationship between our parent black hole and our universe after 13.8 billion years have passed since the Big Bang? The first thing to consider is how much of the interior of our parent black hole is occupied by our universe? If the Big Bang occurred at the innermost core of our young parent black hole, we can assume that we are still well within the "belly" of our black hole. **See the illustration at the beginning of this chapter.** As our parent black hole is expanding, we are chasing the horizon at the edge of the black hole. If we are in a different dimension of space, we will never catch up with this expansion. What may seem a small stretch of space for the expansion of our parent black hole would be unattainable on our own level of expansion within a different dimension of space and time.

Another possible association is that the accretion disk around our parent black hole could be the unknown force affecting the expansion of our universe. The gravitational effect of a growing accretion disk could be the reason our universe is expanding at a much faster rate.

One other connection between our universe and our parent universe is the processed matter (dark matter) that is entering our dimension of space through the event horizon. This unique "dark" matter is processed in such a way that it can enter our universe. Extreme gravity and quantized space can synthesize matter (cosmic soup) at very high temperatures and pressure, producing the "dark" matter that can cross over into our universe. Almost all other matter is destroyed by antimatter as it crosses the event horizon of our parent black hole. Space energy (dark energy) at the outer limits of our universe may also prevent other forms of matter from entering our universe.

In our universe we can see how dark matter is so important for maintaining structure. It appears to be all around us and yet we do not know what it is. It is the life blood of our universe. The reason why it is 27% of our entire universe is because it is everywhere throughout our universe, and it is being augmented as space expands. However, the amount flowing into our universe is less than the force needed to slow the expansion of space toward the event horizon of our parent black hole. We will never see this horizon because it is located in another dimension of space and time.

What we don't know in physics and cosmology is responsible for 95% of the function of the universe we live in! It is easy to accept this fact, if the secrets to the universe are hidden by the event horizon "veil" that holds the answers to how the universe came into being, and how it exists in its present form.

When we try to look back in time through the eyes of technological satellites, such as the Webb telescope, we are able to see a small portion of our parent black hole. We cannot reach back far enough to ever see the horizon of our universe because our horizon is moving away faster than light can travel due to space energy stretching space and time. This would suggest that it is possible for the speed of light to be exceeded. Actually, space energy is stretching space and time. Think of Einstein's illustration of how gravity is a depression in the fabric of space, but now stretch the fabric and the depression becomes elongated. But, since space is stretching in all directions it is elongating space in all directions. Our universe is much larger than we could have ever imagined, and we cannot be sure if it could extend far beyond our current findings.

In our universe the "old" photons we now receive have traveled for over 13.7 billion light years toward the Earth. By the time we receive this light, the universe should have expanded another 13.7 billion light years. However, the Planck satellite and ground-based equipment have suggested that this is not the case. The universe is expanding faster than the standard model. It appears to have expanded an additional 19 billion light years!

Why have we not located dark matter and dark energy in the quantum world using some of the greatest technology humans have created? The Large Hadron Collider continues to search for this elusive matter and energy. The most likely reason for not locating or observing their existence is that they may be as much a part of our parent universe as they are a part of our own universe. They share the characteristics of both universes. When something can exist in multiple dimensions of space and time, it may be invisible to any instrument we might create to confirm its existence.

Another comparison to a black hole is density. How does the density of our universe compare with the density of a massive black hole? The entire mass of our universe located within a black hole that is billion of miles in diameter is not out of the question. It has been estimated that our universe has a mass of 1.5×10^{53} kg. The black hole at the center of the Milky Way has a mass of 8.26×10^{36} kg. The mass of a supermassive black hole is approximately 1×10^{40} kg. One thing that is totally unknown is: What happens to matter when it passes through the event horizon? The newly formed baby universe may not require a close comparison in density since the particle soup reformed into particles within a different dimension of space and time. Also, the matter that we have estimated for our universe exists in our parent black hole of a different universe. In their universe, black holes may be even bigger than the ones in our universe. Thus, the black holes in our parent universe may have masses far greater than in our universe. Maybe our universe exists in a massive black hole that is a trillion times more massive than the largest black holes in our own universe. Another factor that could change the mass of our universe is the flow of dark matter into our expanding universe. Over time this could increase the amount of matter in our universe even though it is spread out over a greater amount of expanding space. Both these suggestions would be enough to significantly alter the mass of our universe. It would be highly unlikely to expect that all universes are the same size. Our universe may be a larger universe, while others may be much smaller. This would be dependent on the size and mass of the parent black hole. Our universe may produce smaller universes than our parent universe. These variables would not require a strong comparison since our universe exists in a different dimension of space.

Most of these massive black holes coalesced between 250 million to 2 billion years after the Big Bang. If massive black holes are the incubators for new universes, it would suggest that there are continuous levels for replication. This could go on infinitely. The separation between generations would be approximately a billion years, however, different dimensions of time and space could alter this time frame by several billion years.

Why couldn't a new universe develop within black holes that are 13 billion years old? This is likely not possible because once they reach an "adult" stage they are too stable to produce offspring. There would be far more anomalies in developing black holes that would encourage the necessary surge of energy to promote the influx of matter toward singularity. This can be compared to the birth of a star when it suddenly starts its fusion process.

The following illustration shows a universe within our parent black hole. Massive black holes in our universe could be a breeding ground for a trillion universes that will have their own black holes to produce trillions more.

Inside our parent Black Hole

As was discussed previously, stellar black holes do not have the mass and structure necessary for this evolutionary process. A specific minimum mass would be required to produce a reaction similar to the Big Bang. This evolutionary process for universe development would suggest that the laws of physics would pass from generation to generation. Just like all types of evolution, there are always changes in patterns that may create different characteristics. But it would likely not change how our quantum world functions. All generations of universes would have quarks, protons, neutrons, and electrons.

At this point, you might pose the question about the possibility whether most universes would evolve with the existence of humans or other intelligent life? Obviously, if the laws of physics transfer between generations of universes, humans should be a part of all the universes. This theory for recreation of a universe seems to make more sense than many similar theories. Other multiverse theories do not provide any clear explanation for creation or evolution. However, these other theories do suggest another topic for consideration. Could more than one universe originate within the same black hole? Considering how nature functions, this could occur. Would this only be possible when two black holes collide, or would it be a common occurrence? As they say, anything is possible. The only issue would be, if two universes exist in an enclosed and expanding space, would the expansion cause the two or more universes to collide? This could be a problem because this circumstance would be comparable to the collision of two large galaxies. In the end, the two universes would form one larger universe with a highly disorganized structure. I could imagine that dark matter would try to recreate some structure in the new larger universe, but this would take billions of years before the new universe could exist without significant chaos.

I called this theory the Black Hole Endoverse Theory because this terminology suggests that a universe exists inside something. In this case, that something is a massive black hole. Why choose a black hole for this theory of evolution? It is the only place in our universe that has the potential to process matter and energy in a way that creates what many call singularity. Singularity has not been proven but the probability for the Big Bang to occur inside a massive black hole could be highly possible under extreme conditions of temperature and gravity. Singularity occurs when the four forces and quantum soup (dark matter) are condensed to a near infinite point. This could be the result of what happens within a massive black hole during its early stage of development.

A good reason for supporting this theory is that it includes the "Big Bang Theory". Why is this important? This theory has been around for almost 100 years. Since 1964, with the discovery of the cosmic microwave background, there has been nothing to disprove this theory. The Black Hole Endoverse Theory simply builds on this established theory, and it provides a solution for infinity.

If there are other universes in the Plan of our Creator, the Endoverse Theory is the most promising method to reproduce our universe or any universe. The theory provides a way to create infinite universes without defying the laws of physics. Too many theories that have been suggested do not provide an answer for how evolution and infinity work together. This is a logical solution to a very complicated problem. A few cosmologists have proposed that black holes are the possible birth place for new universes. However, it is my goal to demonstrate how many intricate processes fit together to establish the probability for proving this theory.

Some have suggested that the quantum world is where other universes exist. For example, could a quark or electron be a universe? This concept totally defies the laws of physics and makes little sense. The quantum world is just that. It is a world in its own right without any connection to being a universe. A universe is not a wave and, or, a particle. One thing we know about our universe, it is not a wave of energy. Quantum particles are the building blocks that make our universe possible.

One of the most significant characteristics of a black hole is spin (rotation). For some reason, supermassive black holes spin slower than smaller galaxy black holes or stellar black holes. Almost all celestial bodies in our universe spin on an axis. If a spinning object's parts are pulled inward, thus increasing density, the object will then spin faster. This is stated in Kepler's Second Law for conservation of angular momentum.

Previously, I discussed neutron stars as the last stage for stars with masses of 2 to 4 solar masses. These stars condense to about the size of a city, consisting of mostly just the nucleus of atoms. All the protons have combined with electrons and the star is essentially composed of only neutrons with little space between them. These dense stars have spin rates that are hard to imagine. For example, the neutron star, PSR J1748-2446 has a spin rate of 716 times per second. This is 25% the speed of light. The density of the star is 50 trillion times the density of lead. This may seem extreme until we focus on black holes. There are black holes in our universe that rotate close to the speed of light. It is estimated that the massive black hole at the center of the Milky Way rotates at an impressive 1150 times per second. Just try to imagine something that is 14.6 million miles in diameter spinning at 1150 times per second. Our Sun rotates 1 time per 27 days. The fastest spinning black hole spins at about 85% the speed of light. The supermassive black holes actually spin slower. This may be the result of how these supermassive black holes formed. Most likely they formed from very large accretion nebulous clouds, and then collided with other black holes early in the evolution of our universe. This would likely produce a slower rate for rotation. Collisions will often counter rotation by cancelling some of the spin from each of the black holes.

These spin rates are determined using the Doppler effect. If the part of the black hole accretion disc is red shifted on the side of the disc moving away and is blue shifted on the other side moving toward, it is possible to compute the speed of rotation. These numbers are still unclear because the accretion disc is not the black hole itself. What is happening inside the accretion disc is not visible since light is not being emitted. We can only assume the speed of rotation would be slightly greater than what was observed at the accretion disc. However, this is an assumption that may be wrong, and there is no way to confirm what we cannot see. A black hole may produce perturbations on surrounding objects, and by supplementing this information with the Doppler effect, it is still not a definite conclusion for the actual rotation.

This discussion about spin rates for black holes was introduced to make a possible connection to a universe inside of a black hole. This can be misunderstood because we are not sure what is occurring inside of a massive black hole. What we have learned from the cosmic microwave background suggests that our universe is not rotating. The entire universe appears to isotropic. Our universe looks nearly the same in all directions. There appears to be no axis and no center. There is no conclusive evidence to discredit the findings from the CMB. If there is evidence that the universe is rotating, it has been suggested that it would have more than one axis. So how does this conflict with the Black Hole Endoverse Theory? We know that most black holes spin on their axis at speeds that are hard to imagine. Yet, our universe has zero spin. Something unusual would need to be considered for this lack of consensus. The question that must be answered: "Can a non-rotating isotropic universe exist inside a black hole spinning at a rate that approaches the speed of light?" Another way to look at this dilemma: "Is it possible for our universe to be isolated from the surrounding chaos of our parent black hole?" There can be only two possible solutions to this challenge. Our universe would need to exist in another dimension of space and time, or space energy would need to be independent from the black hole. Space energy would need to have enough strength to provide a protective layer of resistance from outside influences.

The idea of our universe existing in another dimension is the most likely possibility, when a universe exists within another universe. Space and time were altered when the Big Bang occurred inside the black hole. What goes on outside the black hole is in the background of an additional dimension. The normal activity of the parent black hole will continue as our universe exists as it does today, and into the distant future.

Earlier, I compared the mass of the universe to the mass of a black hole. Now let's compare the size of our universe to the quantum scale that existed during the Big Bang. Our universe is 92 billion light years across (10^{26} meters). If it could be shrunk to the size of a quark (10^{-19} meters) it would be 10^{45} smaller than its current structure. This is the relative difference that closely matches what existed during the process of the Big Bang within a black hole. The current mass of our universe is 1.5×10^{53} kg. Try to imagine this much mass in a space approximately the size of a quark. This is difficult to comprehend, but if you could reduce the size of every particle in the universe by a factor of 10^{45} this would appear within the scope of possibility. In other words, if space and matter were reduced proportionately, all that exists around us today would be just as it is, without altering anything other than space.

With this premise, the Big Bang would not have been a "Big Bang"; it would have been a "Big Whimper". We cannot imagine what happens to space and matter when near the point of singularity. It could be that space just contracts and then recoils rather than exploding. All that really changed was the sudden expansion of space, which propelled the beginning of a new universe. The matter that was necessary to produce this new universe was synthesized by a parent massive black hole in the middle of a galaxy not much different from our own. You may wonder where 1.5 x 10^{53} kg came from if black holes are normally not close to this density. Realistically, the density of the black hole is not that relevant if the new universe that has formed is in a different dimension of space and time.

Earlier, I suggested that the possibility for this evolution of a new universe likely took place in the first part of the development of the parent universe. Most massive black holes coalesced out of huge clouds of gas between 250 million to 1 billion years after the Big Bang, which stirred the pot to help with the formation of large galaxies. During this stage the black holes were extremely chaotic. Once they began to slow down their consumption of these clouds of gas, other stars that had formed nearby began to feel the gravitational effects of the massive black holes. These massive black holes may have collided with other large black holes, and some became supermassive black holes. Dark matter was the driving force for these and other monstrous coalescing structures. As more and more circulation of matter and gas began to collect and form stars, the mass and spin of these black holes created enough gravitational force and turbulence to promote the circulation of gas clouds and stars around its core. This circulation of matter and dark matter grew much greater and eventually galaxies began to form. After the galaxies were formed and the black hole became more like what we see today is when the process of black hole reproduction began. Within the first two billion years of evolution, some of these massive black holes went through a stage where they suddenly had a moment where matter around the event horizon had a sudden burst of energy. These massive black holes started to exhaust huge amounts of ionized material into space, and at the same time this original sudden energy burst pushed a huge amount of their accretion disc through the event horizon. This sudden burst of energy and matter produced a reaction that stimulated the Big Bang at the center of their core. Up to this point, these massive black holes were continuously accumulating matter into their accretion disc. At some point this huge accumulation became so hot and dense that it was like the beginning of the nuclear fusion inside a star, but far more intense. This is the period when black holes first began to exhaust huge amounts of radiation into space. This sudden transformation was the point when the extreme heat and pressure (gravity) at the beginning of this process provided the stimulus for the Big Bang within the sphere of the massive black hole.

The following is a more detailed explanation of how space energy can transform the matter from the event horizon of a black hole and culminate as a new baby universe. This will also reveal what happens before the Big Bang. This event occurs only during the early years of the development of a universe. Imagine a massive black hole when it was approximately one billion years old. Before this point, this black hole was continually absorbing more and more matter into its accretion disk. Finally it reaches a breaking point and the temperature is near one quadrillion degrees Kelvin along the innermost portion of the event horizon. Pressure from extreme gravity and super high temperatures suddenly create a violent surge of energy.

When the force of gravity becomes too extreme, space (dark) energy is no longer able to retain its structure. As a result space begins to contract within the black hole. This action starts the process of pulling almost all the matter from the event horizon of the black hole. Matter suddenly breaks down to become an unstructured quantum soup. Since matter exists within space, contracting space starts to alter the space occupied by the disoriented quantum soup. As space and the quantum soup contract together, space (dark) energy begins to join together the three forces with this quantum soup. Space continues to contract toward a new quantum structure. This contraction stops when gravity and space (dark) energy reach a point of equilibrium. Gravity then consolidates with the other 3 forces and the quantum soup, which has now contracted to the size of a quark. When the four forces and the quantum soup are totally combined, the Big Bang occurs. Gravity, electromagnetic, strong, and weak forces begin to separate immediately after this stage of equilibrium, and the process of the Big Bang proceeds as it was described in the first chapter.

It should be noted that dark matter created along the event horizon transforms to become the quantum soup in this creation process. Normally the production of dark matter along the event horizon is a continuous process that started when the black hole first formed, and therefore dark matter existed even before the Big Bang occurred. Dark matter exists throughout the interior of the black hole which includes the newly created universe. Dark matter will transform into a new unstructured quantum soup only when temperatures and density reach a certain level. Only under extreme specific conditions will this entire process of the Big Bang occur.

For the universe we live in, this event took place inside our parent black hole approximately 13.8 billion years ago. Shortly after the Big Bang, there was a period of time when dark matter was responsible for organizing the hot soup of protons, neutrons, and electrons. A saturated background of dark matter likely provided the stimulus for a faster evolving universe. This made it possible for the early universe to begin forming black holes, stars, and galaxies much sooner than predicted for the standard model. The aggressive involvement of dark matter in the early universe may be the reason the James Webb Telescope is finding black holes, stars, and galaxies that are too well developed in our universe approximately 400 million years after the Big Bang.

It is amazing how our universe can appear so tranquil if we are living within the boundaries of a black hole. This happens only because our universe exists in additional dimensions of space and time. Whatever chaos that might exist is not in our same dimension of space. Our parent black hole served its purpose by providing the environment for the creation of a new universe. Our new universe thrives in its own dimension of space and time unencumbered by the chaos of our parent black hole. At this point, our parent black hole can spin at half the speed of light, gobbling up stars, and thrusting star remnants into space, while we in our own universe have no idea of the surroundings that are hidden in the background of another dimension. The black hole served its purpose for the creation of our universe, and the only things that remind us of our parent black hole are "the laws of physics" that were perpetuated into our universe. All the new universes that have been created in our universe and all universes that were created in the past will be related to each other. Each universe will carry on the traits of each generation for all eternity. Every universe that was ever created will have characteristics similar to our own universe. Every universe should have dark matter, dark energy, black holes, stars, planets like Earth, and of course, the human race.

Chapter 4

Dark Matter
and Dark Energy

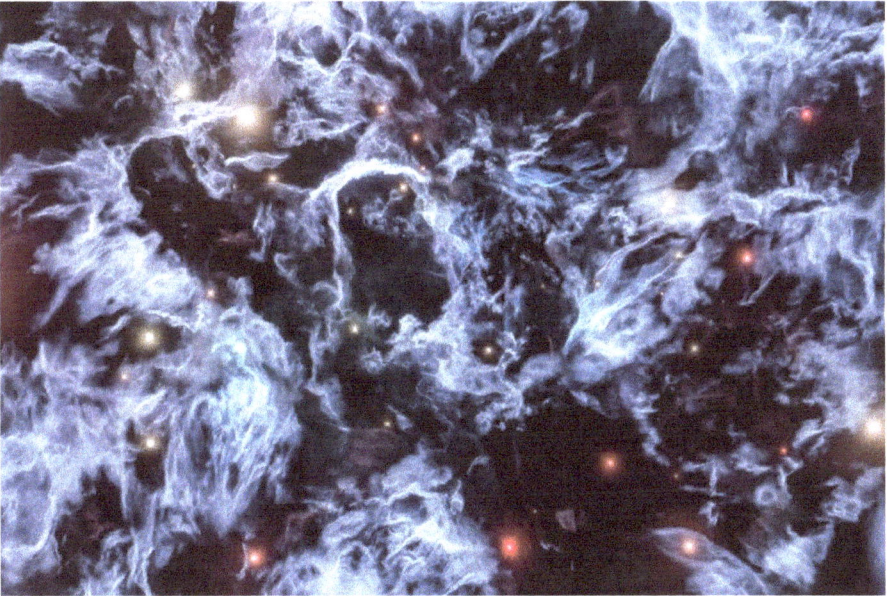

"The missing link in cosmology is the nature of dark matter and dark energy."

Stephen Hawking

Nothing can be more difficult to write about than the unknown. We know so little about 95.1% of our universe. Dark matter and dark energy are almost all that exists and yet we have no clear answers to explain their composition. It is difficult to imagine that standard matter and energy accounts for just 4.9% of the entire universe. Even with our current technology, dark matter and dark energy are presently as much a mystery as they were 25 years ago. I will try to address each topic separately when possible, although they often work together to produce the effects we encounter every day of our lives.

In the previous chapter I suggested that dark matter is produced within a black hole. If this process for the creation of this particle is correct, it will be extremely difficult to understand a particle that is so different than ordinary matter. For this reason it may be very challenging for our current technology to locate this particle.

The quantum world has been literally torn apart in an effort to discover all the particles that make up the universe. Over the past 50 years, with the assistance of particle accelerators, we have confirmed many particles that are now part of the standard model for physics. However, when dark matter is brought into the picture, the laws of physics are greatly challenged. In the end, all the unusual findings concerning dark matter may require a revision of the standard model. The term "Dark Matter" was first used by Swiss astronomer Fritz Zwicky with the California Institute of Technology in 1933. He noticed the Coma Cluster of galaxies did not possess enough mass to hold the structure together. He suggested there must be some unknown matter holding everything together. He called it "Dark Matter".

First let's examine all the suggested possibilities for what dark matter might be. We know that dark matter pervades all space in our universe. Dark matter accounts for 85% of all matter in the universe and yet we have no idea what it is. Is it a particle or simply a force similar to gravity? There are several ways to prove that dark matter exists, whatever its form may be. First, there must be a force greater than gravity that is controlling the movement and order between all galaxies and local groups of galaxies. Gravity alone does not have the ability to organize the pattern that embodies the cosmic web. Secondly, when observing distant galaxies, dark matter has the strength to bend light along the lines of sight of closer galaxies. This is called gravitational lensing. In the quantum world it remains invisible by not acting at all like matter. It does not absorb, reflect, or emit light. The only way to confirm its existence is from observing its gravitational effect on visible matter. Lastly, we know that it has to exist in some form because gravity alone could not have had enough force to cause the coalescing of all the galaxies and stars in our universe. Its existence is so obvious by the pattern it exhibits in the cosmic web of our universe, as illustrated later in this chapter. Its primary reason for existence is that it has the ability to bring structure and organization to our universe. This characteristic may be as important as the structure that exists within the quantum world.

So, what are the possible suggestions for its composition? One of the primary propositions is that dark matter is a weak interacting massive particle (WIMP). This would be a particle that is very heavy, ten to 100 times the mass of a proton, and it would interact weakly with normal matter making it hard to detect. Another suggestion is that dark matter developed in the early stage of the universe as very small primordial black holes. And another possibility, it might be a particle called neutralinos, similar to a neutrino, but slower and heavier. It has also been suggested that it might be a sterile neutrino. These are neutrinos that only interact with other matter via gravity. The most accepted concept among particle physicists is the WIMP proposal. However, so far the LHC (Large Hadron Collider) has found nothing to support any of these ideas. A recent theory proposed by Kathryn Zurek is called the Hidden Valley theory for dark matter. It suggests that dark matter exists as a group of particles totally separate from the standard model.

Only the Hidden Valley theory provides a possible explanation for the existence of dark matter. I say this because all the other suggestions have not been verified by the LHC. Dark matter is definitely more exotic than anything that has been proposed, otherwise physicists would have confirmed its composition by now. Whatever it is, it is far different from normal matter. Accepting this premise, let's look at other options that may explain its true character.

Understanding something mysterious is by definition a difficult pursuit. Most of the time, one has to look at the most obvious fingerprints to obtain a solution. There are two apparent fingerprints that provide the best understanding of dark matter. First, the cosmic web is the most obvious explanation by the pattern it forms to provide structure and stability to our universe. This pattern is so unique that it must be the prime factor for understanding the true nature of dark matter. Second, this structure would not exist if not for dark matter having the ability to help combine all of the primordial matter to form stars and galaxies.

I will begin with the cosmic web. It is difficult to imagine that all the galaxies and groups of galaxies are held together by filaments of hydrogen gas. This spider web appearance gives the sense that this was a design laid out by the best Architect of all time. The enormous scale of this art form is beyond comprehension. This tangled web includes nearly 2 trillion galaxies, each having several hundred billion stars and countless planets. This is the fingerprint of what we call dark matter. So, let us consider what would be necessary to produce a pattern to form the cosmic web. First, it is apparent that whatever it is, it must have enough strength to organize this structure. The shape of the filaments connecting the galaxies suggests that dark matter has an organized mosaic structure. There appears to be a special force supporting the structure of the cosmic web. This would suggest that dark matter is not made of normal particles as has been proposed by others. Also, these connecting filaments are composed of the light gas, hydrogen, so this gas should replicate the true shape of the medium that supports it. Is this web-like pattern what happens to dark matter when it is stretched in its effort to hold the structure of the universe together? This might also suggest that dark matter behaves more like fluid matter, which has the ability to form a shape such as the web-like structure. Dark matter behaves like it is an ocean of water with wave patterns that are extremely organized.

There must be a reason why the universe is held together in this web-like pattern. Recently the Keck Telescope in Hawaii mapped spectrograph images of the cosmic web and this intricate pattern has suggested that all these filaments appear to be pathways for transferring hydrogen gas to galaxies. The entire universe is interconnected by filaments that provide a source of fuel completing a supply chain that not only helped with the evolution of galaxies, but also helps maintain their existence. The pattern for this exotic freeway of connecting galaxies is very similar to the neurons, dendrites, and axons in the brain. **(See Illustrations)** What was discussed in Chapter 2 now becomes even more relevant. Our universe is orchestrated in such an organized pattern that it simply boggles the mind. The immense size of the cosmic web is difficult to describe, if you can imagine this entire web-like structure exists throughout the entire universe.

Dark matter would require the ability to easily consolidate as groups of particles to form a web-like pattern. For greater strength it would suggest it exists as both a particle and a wave. A spider's web is extremely strong, based on the thickness of its strands. This same principle must be a characteristic of dark matter.

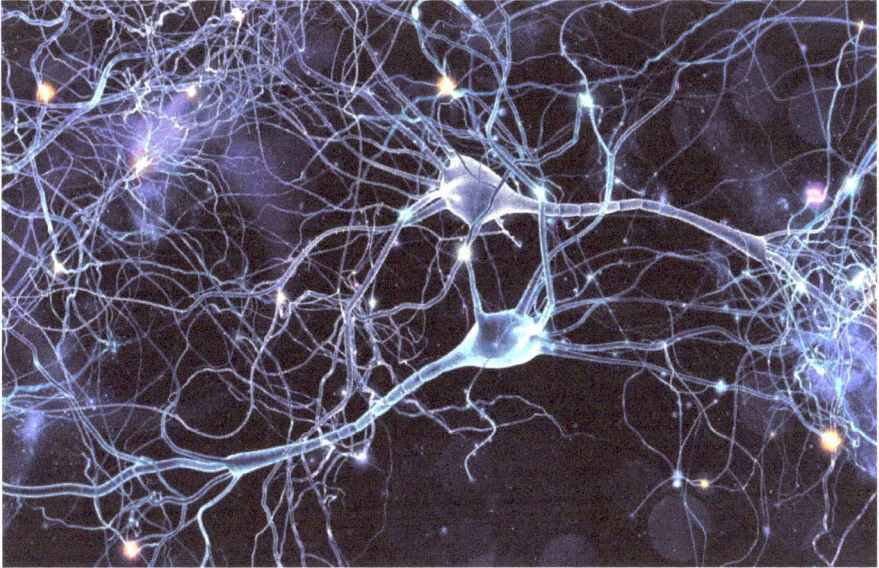

Illustration of Neurons, Dendrites, and Axons within the Human Brain

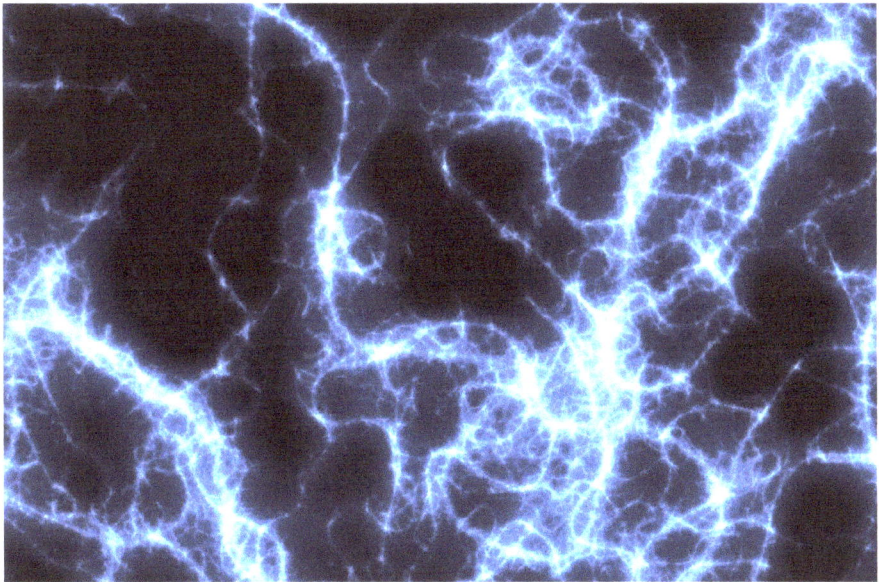

Illustration of the Cosmic Web

We know the filaments of hydrogen do not hold the cosmic web together. It is dark matter that adds the support to accomplish this task. We can conclude that the hydrogen gas is simply mirroring the presence of dark matter. Since this structure exists throughout the universe, we can predict that this pattern was present even during the early universe.

Not only does it hold the galaxies together, it helped pull matter together early in the evolution of our universe. Dark matter had to be very active in the early universe when particles cooled and started to form atoms. During this time, dark matter had to be a dominant force that filled all of space, and it possessed the ability to clump things together. **(See illustration at the beginning of this chapter.)** The illustration represents dark matter when the universe was just beginning to coalesce into stars and galaxies. Without dark matter, all that is present in the universe would not exist. It may be one of the most important ingredients that is responsible for structure in the universe. Without it, there would be no stars, no galaxies, no earth, and no humans. Dark matter's main purpose is to pull or hold things together. Whether it is the coalescing of stars and galaxies, or establishing the structure of the cosmic web, the universe exists because of dark matter.

Earlier I spoke about WIMPs as being one of the prime candidates for resolving the mystery of dark matter. Another preferred choice by particle physicists is axions. They are suggested to be very small mass particles that move much slower than the speed of light. This was also the thought behind the actions of the WIMP particle. Physicists refer to this slower motion as cold dark matter since it appears to react weakly with ordinary matter and electromagnetic radiation.

Two things stand out about the nature of the composition for dark matter. It acts more like gravity with its longer range effects, but it has the strength of electromagnetic waves, but with long range characteristics. This must be an exotic particle different from the norm because of this strong wave tendency. There is something different about how this combination of mass makes it desire to clump ordinary matter together far better than what we observe with normal particles. It is responsible for how it developed the cosmic web in a pattern that seems impossible without possessing some degree of awareness, no different from how a spider can weave its web. Some physicists have coined the Higgs boson as the "God particle". After we discover the true nature of dark matter, I can assure you, the particle that constitutes its existence will deserve equal billing. The manner by which this particle performs should earn the title of "the smart particle". It is smart enough to basically direct the assembly of the entire universe, as if it were following a script. The chaos of black holes producing quasars, the explosion of stars, the evolution of "solar systems", and of course, the cosmic web are all part of the ultimate plan for the evolution of a "smart" universe. All of these components are the result of the actions by a "smart particle" called dark matter.

The axion particle is considered to have a mass that is as small as 1 ten billionth that of an electron. This is likely the case since this particle must exist everywhere in space in order for it to control actions, and even produce shapes and structures in our universe. When we consider that it makes up 85% of all matter in the universe, this makes sense! Even if we accept the fact that the mass of this particle may be small, the force that it exerts is significant. The wave pattern produced by this particle is more like a tsunami rather than an ordinary wave. Basically, dark matter produces the effects of gravity on steroids. Dark matter creates a gravity constant by countering the forces of dark energy. Dark matter is in a battle with dark energy and right now dark energy (the hare) is outperforming the tortoise. In the next chapter I will explain what may be a better solution for the existence of dark matter.

I mentioned earlier that the CMB revealed "cold spots" in one of the hemispheres. Several physicists have even suggested that this could be the result of another universe. If this were the case, what would this nearby universe do to our universe? The problem with this proposal is the fact that these "cold spots" are isolated instances. Therefore, the force of influence (gravity) would be more localized. Since observations show the expansion is occurring in all directions, this would likely eliminate this cause for expansion. Thus, the "cold spots" would need another explanation for their existence.

One possibility for these "cold spots" is the other topic we have been discussing throughout this chapter: dark matter. If dark matter is entering our universe from our parent black hole, these "cold" areas may be a concentration of dark matter. The flow of this dark matter would be fairly even since it would enter our universe in all directions through the event horizon. We know so little about the event horizon of black holes. It is possible that the event horizon of our parent black hole may have areas that are influenced by magnetic forces and lower temperature spots, not that different than sunspots on our Sun. The "cold spots" could be a deeper impression of what might exist at the event horizon of our parent black hole, or they may be the remnants of the early stages of our evolving universe where concentrations of dark matter may have existed.

In chapter 2, I mentioned two main reasons for the aggressive behavior of dark energy. First, it is the nature of space to expand. Second was the outside influence of another universe. Although the most likely possibility is that space just wants to expand, the influence of an outside force cannot be overlooked. And, of course, both of these possibilities could be working together to create the aggressive behavior we now observe.

Previously, I suggested that dark energy is really "space energy". When dark energy is given this name, it is easier to describe how it works. I will be using both titles interchangeably, but they are the same. It's just that everyone is familiar with the term, dark energy. If dark energy is considered to be space itself, it becomes easier to explain the events that are taking place in our universe. Runaway inflation is a real problem for our universe. Given enough time, it will tear apart all that seems ordinary in our universe. This expansion of space is beginning to affect the cosmic web, which is the life blood for the galaxies in our universe. Dark energy is literally pulling apart the filaments that connect our galaxies. So, what is causing this rapid expansion? And, will it ever begin to slow down?

Dark energy (space energy) appears to have one established trait: It wants to expand. This inherent quality is all part of what started with the Big Bang and it continues to this day. But something happened about 5 billion years ago to cause the universe to begin expanding faster than the normal progressive expansion. If we want to know the future of our expanding universe, we have to look at what must have taken place 5 billion years ago. What changed to promote this more aggressive expansion? This time period would be about the time our Sun and solar system were forming into the structure we see today. What could have been happening during this era?

Most of the findings about our past universe are revealed by studying the Cosmic Microwave Background. By analyzing the temperature variations on this cosmic map we can determine whether something is isotropic or there are fluctuations that have taken place in the structure of the cosmos. Expansion shows something being redshifted when photons emitted have decreased energy. This is difficult to see on the CMB. But if comparisons are made with satellite and terrestrial observations, the changes in redshift become apparent.

Something unexpected must have occurred 5 billion years ago to promote the aggressive form of expansion. As I stated earlier, expansion may be more the battle between dark matter and dark energy, rather than some major catastrophe that may have occurred. The effects of rapid expansion may just be the result of the aging of the universe. It may be the natural process of what happens to this or any universe. In the previous chapter, I discussed the evolution process for the creation of new universes. This may be the plan for how universe(s) evolve. Everything around us goes through stages of development and then eventually begins to reverse course. Even though many things in our universe die, they often are replaced with new creations. Stars are constantly dying and are replaced by new stars. Life on our planet follows this same pattern of evolution. It is just part of a bigger Plan.

Finally, I must refer back to the previous chapter for what may be the best explanation of excessive expansion of our universe. This expansion may be the influence of gravity from our parent black hole. The gravitational effects of the huge accretion of matter surrounding our parent black hole could permeate our universe even though it exists in another dimension. We do not understand gravity well enough to know whether this is possible. In other words, can gravity cross different dimensions of space and time to influence space energy (dark energy) in our universe? The growing accretion disk of our parent black hole may be the main reason our universe is expanding at rates that exceed our predictions. Could this be the answer to solving this mystery of excessive expansion?

Astrophysicists are now suggesting that our universe may extend 46 billion light years from any point, making the diameter 92 billion light years across. If our universe is believed to be 13.8 billion years old, you can see the discrepancy must be the result of expansion exceeding the speed of light. We now refer to approximately 13.8 billion light years as the "observable universe". The stretching of space makes it possible for the speed of light to be exceeded. Because of this fact, we cannot be certain just how far our universe may extend. Is it possible the universe is larger than 92 billion light years? Yes, because we are merely guessing beyond the known universe what is really taking place. We base these estimates for expansion on what we observe and then assume this proportionality continues into the invisible undefined universe.

Due to expansion, the future for our universe may seem bleak. Is it possible that space energy could weaken and allow dark matter to regain its previous status? At this point, this appears unlikely. But the good news is that, it will likely be more than a trillion years before everything turns out to be lifeless and dark. Maybe one trillion years from now dark energy will live up to its name. That would be the time to call it "dark" energy.

Chapter 5

Quantum World

"For us physicists, the distinction between past, present, and future is only an illusion."

Albert Einstein

One thing that is common to space, it is basically empty, even within the quantum world. For example, the hydrogen atom is 99.9999999999996% empty space. The universe is 99.9999999999999958% empty space. And, the universe becomes "more" empty by combining the space within the universe and hydrogen atoms. To try to comprehend the empty space inside the hydrogen atom it would require a little more imagination. Everything in an atom is more diffused. Electrons are very spread out, appearing more as a cloud filling some of this empty space. Most of the empty space in atoms is filled with forces of energy, such as electromagnetic waves and forces of energy produced by gluons. But, overall, most of an atom is empty space, if we specifically consider the presence of mass.

Even our own human body is 99.9999999% empty space. The human body has 7×10^{27} atoms, which is mainly just space. What is amazing is that everything we know and see is primarily empty space. Even though empty space makes up matter, a table is still a table, our earth is the place we live, and we are who we are. How can objects exist with so much empty space? I will provide a solution to this question at the end of this chapter.

Just how fast does the electron orbit around the nucleus of an atom? It travels at $1/137^{th}$ the speed of light, which would be 2.2 million meters per second. The circumference of a hydrogen atom is 5×10^{-15} meters. This would mean that the electron mass orbits the proton nucleus 10^{23} times every second. That is 100 billion trillion times per second! Since the electron is a particle and a wave, this creates what is called the "uncertainty principle", first proposed by Werner Heisenberg in 1927. Because the electron behaves as a particle and a wave, its position and movement cannot be specifically confirmed or predicted. The illustration at the beginning of this chapter shows the blurred path of electrons orbiting a nucleus. To be more accurate, the orbits should be represented by a larger wave pattern, meaning the orbits would appear more scattered and diffused (cloud), thus reflecting a greater uncertainty for the position and movement of the electron.

A proton is 1,836 times more massive than an electron. But the main reason an electron orbits a proton is the difference in electric charge. The positively charged proton attracts the negatively charged electron with just the right charge to keep its smaller mass in a specific orbit. The number of protons and electrons are the same in neutral atoms. In this state, the particles are said to have a balanced charge. If energy is gained, atoms can acquire electrons from other atoms. When energy is lost, atoms may lose electrons. When atoms gain or lose electrons, they are said to be in an ionized state. Also, when the charge in the nucleus becomes stronger (more protons and neutrons) there is a stronger attraction, so the radii of orbits become smaller. The atom does grow in size as more orbits are added, but not proportionally. As an example, the calcium atom has 20 protons, 20 neutrons, and 20 electrons. The electrons are positioned in 4 orbits. The first orbit has 2 electrons. The second has 8. The third has 8, and the fourth has 2. Just because there are four orbits does not mean the entire atom is four times bigger than the hydrogen atom. When the electric force is greater, the radii of the levels are smaller. The radii of atoms vary. In some cases, higher numbers on the periodic table are not always larger as predicted. But generally, more orbits equal a slightly larger radius. The radii variation is normally determined by which level has the most electrons. The outer orbit normally does not have the largest number of electrons. This is because it is further from the nucleus, and as a result, is more likely to become ionized. A good example of ionization is electricity. Copper is a metal that is a good conductor of electricity because it has only one electron in its outer shell. The electromagnetic force on this outer orbit is also less, making the exchange of electrons much easier when energy is introduced. When copper wire exchanges electrons from one atom to the next, electricity is produced. All metals have one, two, or three electrons in their outer shell. Aluminum has been used for electricity, but it does not conduct as well. It has three electrons in its outer shell. Copper is also better because it is less likely to oxidize, which can slow the flow of electricity.

There are 118 elements that make up the periodic table. Dmitri Mendeleev started this table in 1869. It was first discovered that higher level shells normally hold more electrons. For example, the maximum number of electrons in the first orbit is 2. The second is 8. The third is 18. There are a total of 7 shells (orbits) in the heavier elements.

After reviewing the basics of the atom, it is time to look at the true nature of the quantum world. There are 17 fundamental particles in the standard model. The main particles are quarks, electrons, and photons. These three particles make up almost the entire universe. So, why are there other particles?

MATTER
from molecule to quark

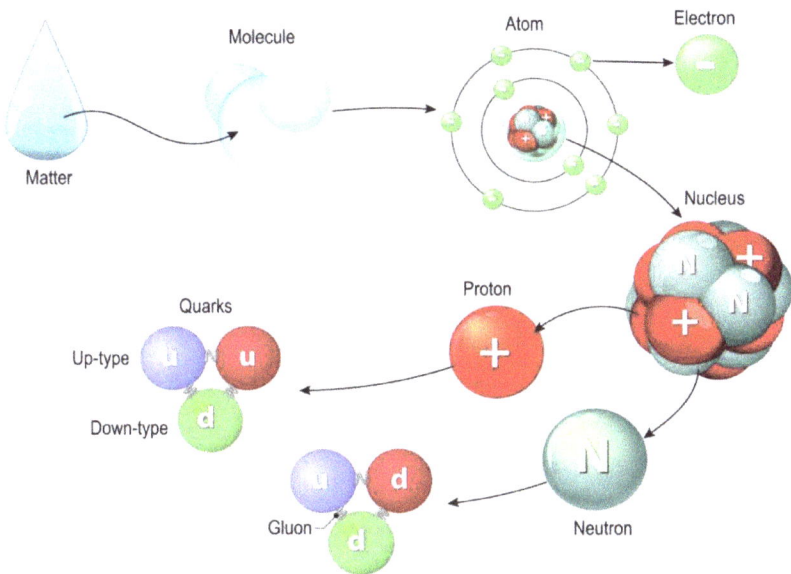

Quarks are thought to be the smallest form of matter within the atom. There has been some debate whether there is still a smaller particle that makes up the quark. The preon is still just a hypothetical particle. At this time, there have been no findings to support the existence of a particle smaller than the quark. Outside the atom, there are particles much smaller than the quark. This would include neutrinos, gravitons, and another particle I will discuss later in this chapter.

The proton and neutron are each composed of three quarks. With particle colliders, physicists have observed three minimally defined particle trails produced by proton collisions. The strong force within the nucleus prevents clear lines for visualizing quarks. By coupling spin and charge with these faint trail signatures, physicists can confirm the existence of quarks. Particle physicists have verified there are 6 types of quarks: Up, Charm, Down, Bottom, Top, and Strange. Protons consist of two UP and one DOWN quark. Neutrons consist of two DOWN and one UP quark. UP quarks have +2/3 charge and DOWN quarks have -1/3 charge. The other four quarks discovered by the LHC are all heavier than the up and down quarks. These have life spans in the billionths or trillionths of a second, often transforming into different types of quarks. Essentially, the up and down quarks are what we know as matter. What is interesting is that quarks are 98% energy and 2% mass. Quarks create this heavier mass in accordance with Einstein's equation, $E = mc^2$. Protons are actually 100 times heavier than the 3 quarks. It is the fast churning movement of the quark that increases mass to match what we see in the proton.

Quarks are 2000 times smaller than a proton. So what fills the extra space between the three quarks that constitute the proton? The three quarks are bound together by gluons. Gluons hold the churning quarks in the space known as the proton. Gluons are responsible for what is called the strong nuclear force. Just like gravitons, gluons may also behave like strings of energy.

All the remaining particles have been discovered by smashing particles moving at near the speed of light. Most of the recent discoveries have been accomplished at CERN or Fermilab. CERN stands for Conseil Europeen pour la Recherche Nucleaire or European Organization of Nuclear Research. The idea for CERN was first proposed in 1949 by French theorist Louis de Broglie. In 1954, 12 European countries established CERN as the largest physics laboratory in the world, and determined it would be located in Meyrin, a suburb of Geneva, Switzerland. Today there are over 2,500 employees and 12,000 institutions working together to collaborate information. More than 70 countries are now involved in this scientific endeavor.

In 1957, the 600 Mev Sychrocyclotron became operational. Physicists first focused on analyzing the atomic nuclei, but soon broadened their focus on high energy physics. This machine had a 52-foot circumference, a far cry from what was being planned. The next level for quantum research was the Large Electron Positron Collider completed in 1989. The desire was to accelerate particles nearer the speed of light. To accomplish this goal, they dug a circular tunnel 27 kilometers (17 miles) long, roughly 165 to 575 feet underground. It took three years to dig the tunnel. It included 5,176 magnets and over 100 accelerating cavities. They also had four different detectors for recording the impact of high speed particles. Even with this new collider, a more advanced Large Hadron Collider was in the planning stages. In 1995, a proposal was accepted by CERN at an estimated cost of 4.6 billion CHF (SwFr). Construction began in 1998 and the LHC was completed in 2008. They were able to use the previously constructed tunnel system with highly updated equipment.

The LHC served its purpose by discovering one of the most important pieces of the quantum puzzle. In 2012, the Higgs boson particle was discovered. It was almost 40 years after this particle was proposed by Peter Higgs. The scientific community went crazy over this long anticipated discovery. It was difficult to discover because heavier subatomic particles decay quickly into photons of energy.

Its significance is important because the Higgs particles within the Higgs field are responsible for giving mass to quarks and electrons. The Higgs boson particle has a very short life span (less than a trillionth of a second), but it accomplishes a very important role, giving mass to matter.

Many of these strange particles have very short life spans. They pop in and out of existence, but during these billionths of a second they serve a specific purpose, and then transform into another form of matter or energy. Even though their life spans are short, the force they create is mandatory for how matter and energy function. Dark matter and dark energy may be partially responsible for this strange characteristic of how particles suddenly appear out of background forces of energy.

Two other important particles are the W and Z bosons. These particles work together to produce the weak force. The gluons created the strong force to hold the nucleus of the atom together, while the weak force is necessary for transforming atoms so fusion will occur inside stars. The W and Z bosons produce electric charge that changes the flavor of a quark, which causes hydrogen protons to fuse, thus generating the fusion process. Without this important process stars and galaxies would not exist.

Another reason why quantum physics can be very confusing is because there are too many titles. Some of the particles are separated by categories and unfortunately this adds more titles into the mix. The categories include the following: Hadron particles include quarks (protons and neutrons), gluons, and antiquarks. This is broken down further with two more titles: baryons and mesons. Baryons are composed of any three quarks. Mesons are a combination of a quark and an antiquark. Baryon particles are part of the building blocks for the existence of matter. Mesons do not appear to have a specific purpose, but they do allow physicists to better observe interactions between quarks.

Another category is Leptons. These are elementary particles that include electrons, muons, tau, and neutrinos. Whereas quarks are social particles, leptons prefer solitary status. Muons and tau particles exist for billionths of a second and have little or no purpose. Just like mesons, they provide better observation for particle reactions. Neutrinos, on the other hand, are part of the nuclear fusion process. They appear to stimulate radioactive decay. Neutrinos are stable particles, meaning they do not decay like many of the other particles. They are also very small particles. A quark has a rest mass of 2.4 MeV, whereas a neutrino has a rest mass of 1eV. For years, scientists have tried to capture neutrinos with no success. They are invisible and can pass through anything including lead. They pass through our bodies all the time. Although they exist almost everywhere they cannot be seen. The mystery behind neutrinos is similar to that of dark matter. A few physicists have even suggested that neutrinos may be part of dark matter.

The quantum world is a place that Einstein called "spooky". There can be no disagreement for this assessment. The function of particles in the atom is more about probability than reality. Once scientists have a clearer understanding of dark matter, some of the spookiness may be eliminated. Since dark matter comprises 85% of all matter in the universe, it has to be part of this crazy quantum world. I still find it so difficult to understand why dark matter has not revealed itself in any test physicists have conducted. It is as if dark matter comes from a background universe different from our own. I mentioned this possibility in the Black Hole Endoverse Theory. It moves around freely throughout space and does not interact with other matter, except gravitationally. It appears to exist everywhere in our universe, but we can't see it!

Is it possible that dark matter may be something totally different? Could dark matter simply be a form of gravity with mass? What if there a different type of gravity that is not dependent on mass? It could be influenced by mass and sometimes it would not be influenced at all. Astronomers have discovered galaxies that appear to have a lot of dark matter, and then observed others that appeared to have no dark matter present in their surroundings. Why the variation? Does dark matter collect in just certain areas of space? These gravitational variations do not exist with normal matter. This variable suggests that dark matter acts like gravity with its own mass.

If dark matter is an exotic particle, what special characteristics would it possess? What would happen to gravity if it had mass? First, it would be capable of attracting it own mass, which would encourage particles to group together. This added strength for gravity would be what is necessary for stars and galaxies to coalesce. Since dark matter constitutes 27% of our universe versus just 5% for all ordinary matter, one can conclude that it must exist appreciably throughout the universe. One possibility for why there is so much dark matter could be what I proposed earlier; this exotic form of gravity and matter is continually flowing into our universe. Our parent black hole is creating the dark matter that exists within our own universe. Because this exotic particle is from a universe beyond our own, it will possess special traits. First, it would have mass since it is part of the soup of matter that enters our universe through the envelope of the event horizon. As discussed earlier in the Endoverse chapter, the only way to cross this horizon is for matter to be totally re-created. During the Big Bang all particles entering our parent black hole horizon would be reconstructed or annihilated as density and temperatures approached a point near singularity. The only reason that this occurs only early in the development of the black hole is because the density and temperatures were great enough to cause dark energy (space energy) to collapse. Gravity overcame space. The current condition in our parent black hole is not as extreme as was the case during the birth of our universe. Since gravity is not strong enough to overpower space, a smaller amount of this exotic particle is flowing into our universe. This exotic particle is what we call dark matter. It was flowing into the core of our parent black hole even before the Big Bang occurred. Since this flow of dark matter has been constant from the beginning of the universe, it would not reveal itself in the CMB map. There are some areas in the CMB that suggest "cold spots" which cannot be explained. This could be where there was once a concentration of dark matter that deviated from the normal flow into our universe. There have been recent observations suggesting that dark matter is not isotropic.

Dark matter has properties that suggest its structure is different than a graviton. Gravitons move at the speed of light, whereas dark matter will move slower because it possesses mass. If it was processed in a different manner, as would be the case, it may be very difficult to discover in the quantum world. Most likely it will be infinitesimally small, making it even more likely to escape detection. With mass it will facilitate the coalescing of matter. It has the ability to cause gravitons and matter to "clump" together which acts like a force field to enhance the effects of gravity. The following illustrations show how gravity supplies a force to secure structure in our universe. Dark matter and gravitons work together to produce gravity.

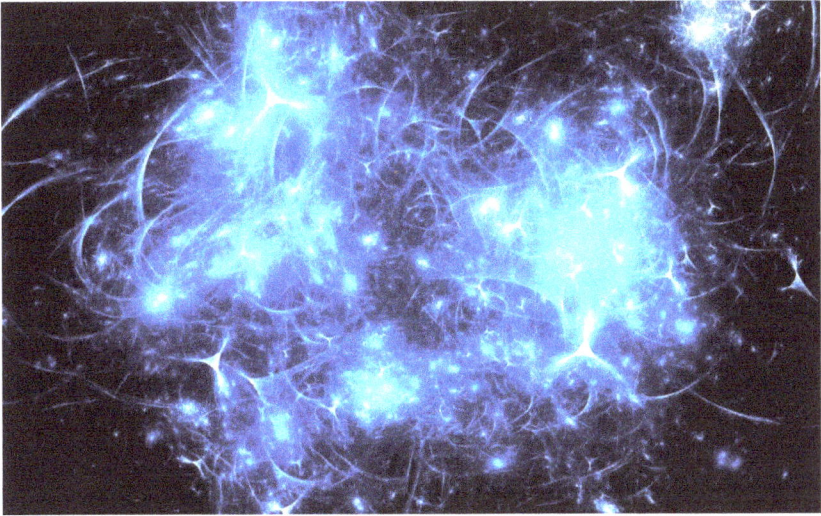

This illustration represents gravitons when there is no mass nearby. Gravitons are disorganized strings of potential energy unless mass is present.

This shows a dark matter force field surrounding the Earth. Dark matter assists string-shaped gravitons in producing gravity throughout the universe.

The first illustration portrays gravitons in a disorganized chaotic pattern when there is no mass nearby. When a body of mass is present the string-shaped gravitons are additionally attracted to dark matter, which produces a force field near the body of mass. This combination of gravity is the force needed to keep our planets in orbit around the sun. Dark matter acts like a version of gravity, but also has mass to produce the needed additional force for keeping our universe in order. Just as gluons are responsible for the consolidated structure of the proton, dark matter is the "glue" that helps hold everything together on a larger scale.

So how does gravity actually work? Einstein's theory of relativity suggests that space is warped by the existence of mass. This stretching of space will act like gravity by attracting nearby matter. This may be a great explanation for gravity, but recent research by LIGO (Laser Interferometer Gravitational-wave Observatory) has discovered that gravitational waves are produced when violent collisions of neutron stars occur. These same studies verified that gravity waves travel at the speed of light. Gravity waves are essentially proof for the existence of "gravitons". Therefore, gravity must still include Isaac Newton's Laws as well. This would suggest that gravity follows the thinking of both of these great minds. It is this combination of gravitons and dark matter that is needed to produce a greater force. Earlier, I discussed how the universe is almost empty space. For gravity to function over such great distances, it would require warped space, gravitons, and dark matter. All three ingredients are required to produce the forces needed for the structure of the universe.

The following illustration shows how string-like gravitons function to produce gravity. The force of gravity is produced only when gravitons align and connect to each other. These graviton chains can link together over great distances, but the connections become fewer as the distance increases. The force of gravity is determined by the number of connections and the distance from the origin of the mass. Gravitons do not have mass. Gravitons travel and connect at the speed of light.

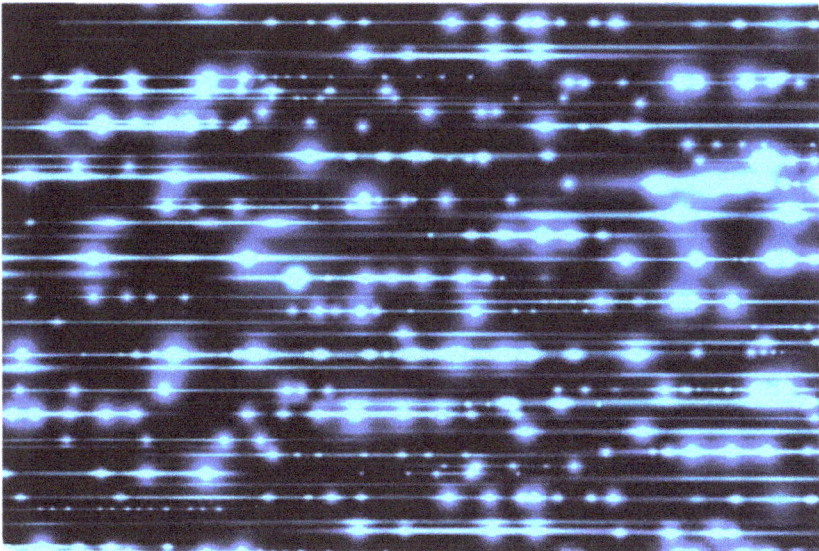

This illustration provides a close-up of gravitons making continuous connections to produce the force of gravity. Gravitons transform from being meandering string particles to organized lines of gravitational force.

This graviton transformation activity is comparable to the brain involving how connections are made between neurons. This overall proposition makes sense because particles with a string-like shape would be conducive for making continuous connections for achieving a force of gravity over great distances. It will be difficult to locate gravitons in the quantum world because they are basically inactive, since there is so little mass in the atom.

Assisting in the process of producing gravity is dark matter. In the previous chapter, I provided many suggestions that physicists have proposed for the resolution of what dark matter might be. If we accept certain elements as fact, then a conclusion might be possible for resolving the composition of dark matter. We know that dark matter has a strong gravitational effect on visible matter. We know it can cause matter to coalesce. It appears to contain mass from the effects of gravitational lensing and the added gravity needed for organized patterns in galaxies. We know it does not appear to interact with ordinary matter except through gravity. And, if we can accept the proposition that dark matter was created within a black hole, we can now answer the question: What is Dark Matter?

Dark matter is an extremely small particle. A neutrino is a million times smaller than an electron. This particle would be a billion times smaller than a neutrino. Dark matter particles should fill much of the void of space. The perception of dark matter could best be described as the "ocean" of space. Since dark matter can exist in multiple dimensions of space, it must be considered an exotic particle. Even though the mass of this particle may be small, it will often combine into groups of particles and exert a force in a wavelike pattern. These groups can combine to form even larger structures, which becomes a large force field that can direct gravitons and increase the force of gravity. Its ability to consolidate gravitons will produce very strong waves of gravitational energy. As discussed earlier, this grouping characteristic is revealed in the structure of the cosmic web. Since all space possesses so much of this cloud of exotic matter, it amplifies gravity throughout our universe. Its existence is as significant as space and time. Without this controlling force, nothing could exist. **See the illustration** of our universe inside our parent black hole on page 61. Imagine the foreground of blue wave bands as the "ocean" of dark matter that exists throughout the core of the black hole, including our universe.

Dark matter is the "life-blood" of the universe. If we conclude that this exotic matter is from another universe, Fritz Zwicky coined the appropriate name for this particle. Dark matter was produced by our parent black hole early after its formation. Dark matter was already present within the entire interior of our parent black hole even before our universe was born. Before all the particles we know today settled into their existence after the Big Bang, dark matter was already in the background of space. This is the main reason why our universe evolved so quickly from the very beginning.

Previously, I suggested dark matter will be difficult to discover because it is very different than other forms of matter. This is because it can exist in multiple dimensions of space. Massive black holes may be one of the most important pieces of the cosmic puzzle. They are the birthplace of new universes and they produce one of the most important particles for the existence of the universe: Dark Matter.

At the beginning of this chapter, I discussed how everything in the universe is so empty. What may appear empty can be deceiving because of one thing – dark matter. Empty space may not be quite as empty as we think. Hidden in the background of space is "black hole" dark matter that fills the space within and beyond our universe. Dark matter is 27% of the entire universe. Space is not empty!

It is difficult to imagine how large celestial bodies can move through the immense void of space and remain in such an organized configuration. Dark matter is what orchestrates this phenomenon. Since most of the dark matter we know was created inside a black hole, it would be more relevant to rename this particle "Black Hole Matter".

If these theories concerning dark matter are proven to be true, it would verify that God created the universe in such a way that it would last for all eternity. All the pieces of this complicated puzzle will begin to fit together if "dark matter" is created in a black hole.

Chapter 6

Human Significance

"We are the representatives of the cosmos; we are an example of what hydrogen atoms can do, given 15 billion years of cosmic evolution."

Carl Sagan

Our journey has taken us from the smallest known particle to the edge of our universe. We did not venture further in the attempt to discover what may lie beyond these boundaries. For that matter, there may be no boundaries, where space and matter extend forever. What lies beyond will never be known except by the Creator of this complex universe. It is very apparent that the universe is far more complicated than scientists could have even imagined. The real question that we all have contemplated: What is our significance in this immense universe, that stretches 92 billion light years, filled with two trillion galaxies, each with over 200 billion stars, and countless planets? The significance of Earth is comparable to specifically attempting to locate one grain of sand hidden somewhere among all the beaches on our planet. Why create such a vast universe for the human race?

All that has been discussed thus far does not compare to the complexity of trying to understand our purpose within this prodigious volume of space. What is even more humbling, the universe we know may be just a small portion of an infinite number of universes. In the first chapter I spoke about whether or not a tree made a sound when it fell, if there were no one to hear it fall? The answer as stated by George Berkeley, the Anglican Bishop: "Yes, it will make a sound, because God will hear it." This thought should be considered when we try to understand how we fit into a plan for our significance. Even before humans existed, the universe had a purpose and it did exist without our acknowledgement. All the components of the universe fit together and culminate in the Plan for replication and infinite existence. So why did God add humanity to the ultimate Plan for the universe?

If intelligent life was prevalent on a large scale throughout the universe, it would be more obvious that God had a purpose for our existence. There should be an abundance of intelligent life considering the enormous number of stars and planets. Combining all the planets and stars in our universe would be equivalent to a trillion trillion. This huge number definitely improves the odds for the existence of many life forms similar to us. If we consider our galaxy to be a standard, how prevalent is intelligent life?

Our Sun is an average star with 8 planets. Most of the stars in our galaxy should also have several planets. How many of these stars could have planets capable of supporting life? If we can analyze all the types of stars, we may be able to determine the abundance of intelligent life in our galaxy. Let's begin with red dwarfs, which are the most common stars in the Milky Way. Most red dwarf stars are one half the diameter of our Sun or less. Many should have 2 or 3 orbiting planets. However, stars this small are less likely to support life. The red dwarfs that could support life would likely be the larger stars of this classification. These stars have a greater lifespan that is more than 10 times a star like our Sun. A few very small red dwarfs may survive for over a trillion years. Their added lifespan may offer a better chance for life to evolve, but even the larger red dwarfs have many challenges that could prevent life from developing within their planetary system.

Our nearest star, Proxima Centauri, is a red dwarf star. Of the sixty nearest stars to our Sun, fifty are red dwarfs. Three quarters of all the stars in our galaxy are red dwarf stars. In comparison, stars like our Sun make up just 8%. Still, this figure would suggest that there are over 25 billion suns (stars similar to our Sun) in our galaxy.

The main problem for red dwarfs to have a habitable planet is low surface temperature. They average 6,000 degrees Fahrenheit versus our Sun at 10,000 degrees. For a planet in a red dwarf system to be in a habitable zone, it would need to be much closer than the Earth to the Sun. Their habitable zone would need to be between .05 to .1 Astronomical Units. The Earth is 93,000,000 miles (1 AU) from the Sun. That would require a planet to be 5 million to 10 million miles from an average red dwarf.

Mercury, the closest planet in our system, is 36 million miles from the Sun. Even at this distance it has surface temperatures near 800 degrees Fahrenheit. At 10 million miles, the average red dwarf might appear in the sky 4 times bigger than our Sun, but it would not produce any more heat at this distance. In addition, red dwarfs would have fewer planets since they coalesced out of a smaller cloud of gas and dust. This reduces the chance for planets to form within this limited environment. Another problem is tidal locking. Just like our Moon, a planet that keeps the same surface facing the star at all times would have extreme surface temperatures. Tidal locking is common when planets or moons are close to their primary planet or sun. Another problem is that stellar fluctuations are more prominent when the star is closer to the planet. Still, these problems do not rule out the possibility for life to evolve in a red dwarf system. As I mentioned, it would likely need to be the larger red dwarf stars for life to evolve. However, for intelligent life to evolve may be nearly impossible because of the reasons I just mentioned. The Goldilocks zone is far too close in a red dwarf system.

Stars are classified into groups based on surface temperature. These spectral classes of stars are identified with letters in our alphabet, starting with the hottest stars. The letters are: O, B, A, F, G, K, and M. When I was in college we used a phrase to remember these classes. Maybe it was chauvinistic: "Oh Be A Fine Girl, Kiss Me". Red dwarf stars are the coolest. Blue stars are the hottest. **Shown below is a list of stars based on class:**

CLASS	COLOR	TEMP F	*LIFE span (yrs)	MASS no. solar	RADIUS no. solar	%STARS main sequence
O	Blue	50,000-100,000	3 million	64	16	0.00003%
B	Bl-White	18,000-50,000	10 million	18	7	0.13%
A	White	13,000-18,000	600 million	3.1	2	0.6%
F	White	10,000-13,000	3 billion	1.7	1.4	3%
G(Sun)	Yellow	8,500-10,000	10 billion	1.0	1.0	7.6%
K	Orange	6,000-8,500	25 billion	0.8	0.9	12.1%
M	Red	3,000-6,000	100 billion	0.4	0.5	76.45%

*Some figures are approximates since the figures reflect a range.

When reviewing the classes of stars shown in the chart, it is apparent that "K" and "G" class stars provide the most likely possibility for life to evolve. Their temperature is similar to that of our Sun. Also, their life span provides a better opportunity for life to evolve on a planet that would be in the habitable zone. The Sun has a classification of G-2. The "K" and "G" stars make up almost 20% of all the stars in our galaxy. That means there are approximately 60 billion stars in our galaxy that could be possible candidates for life. Based on how these stars evolve, it is very likely that the vast majority of these stars have planets. The question is, how many of these stars have terrestrial planets? And then, how many of these planets are at just the right distance from their home star to be able to support life?

If "K" and "G" stars make up 60 billion stars in our galaxy, and if each of these stars possessed 4 to 5 planets, the number of planets for these star classifications could be 300 billion. If just one third of these planets were terrestrial planets, the number drops to 100 billion. If just one out of ten of these planets had water and other elements necessary for life to evolve, and if just one out of ten of these planets were in the habitable zone, the number would still be one billion planets in our galaxy capable of supporting life. These numbers cannot be that unrealistic, so the chances for life to exist elsewhere in our galaxy should be quite high.

Of the 6,000 plus exoplanets that have been discovered around other stars, many have been found to possess water. Many of the water worlds we have found are much bigger than Earth. This is because it is harder to locate smaller planets similar to Earth. We are still in our infancy in discovering exoplanets. So far, most of the nearby stars we have studied have more than one planet. It is difficult to determine if most stars have as many planets as our Sun. Most of the planets we now discover are only located by how the primary star wobbles around its center mass, or by how the luminosity of a star may vary when something transits across its surface. Approximately 70 planets have been discovered by direct imaging. These were larger planets and most are not terrestrial planets. Until we have a better method for locating and analyzing exoplanets we will only be guessing the number of planets that are out there.

Until recently, the estimates for terrestrial planets have been greatly underestimated. I say this for two reasons. First, I just mentioned one, which is not being able to study the wobble of a star well enough to discover planets as small as the Earth. Secondly, the formation of planets around stars similar to our Sun is likely more common than we may have considered. It is very probable that a class "K" or "G" star could not coalesce unless the accretion disk is inordinately large and dense with gasses from previous exploding stars. These stars should possess more than 4 to 5 planets. Stars like our Sun are at least third generation stars formed from larger clouds, making planet formation more common. If this commonality for coalescing can be confirmed, it will be accepted that there may be far more planets in our galaxy than we could ever imagine.

To determine if stars could possess planets in the habitable zone, it is important to establish if their luminosity is safe for life to flourish. Since the Sun is ideal for intelligent life, it is used as the standard. There is a relationship between luminosity and the temperature of the star's radiating surface. Our Sun has a surface temperature of approximately 10,000 degrees Fahrenheit. To determine luminosity for a star, it is dependent on mass. For example, if a star's mass is 2 times that of our Sun, the luminosity will be 2 to the power of 3.5 ($2^{3.5}$). Therefore, this star would be 11 times more luminous than our Sun: This also affects the life span of the star, which is: 2 solar mass/11, or 0.18 the life of the Sun. If our Sun has a life span of 10 billion years, then a star 2 times our Sun's mass will run out of hydrogen fuel in just 1.8 billion years. So a star that is just double the mass of our Sun would have too short of a life span to provide any possibility for life to evolve. This is why it is important to understand how the mass of a star will affect the search for life in our galaxy. Basically, if a star has a life span less than 4 billion years, the possibility for intelligent life to evolve is very small.

Another example, an orange dwarf star with 0.8 the mass of our Sun, would compute as follows: $0.8^{3.5} = 0.46$ the luminosity of the Sun. This star would live 17 billion years: 0.8 mass/0.46 = 1.7 times the life of our Sun, which is 1.7 x 10 billion.

The Earth is located in the Goldilocks zone. This zone for life is very thin. Our habitable zone is .9 to 1.2 AU. Life on Earth would not exist if the Earth were just 8 million miles closer or 18 million miles further from the Sun. This variation is approximately 10%. This is a very fine line for our existence. This is likely true for any civilization that may exist somewhere in our galaxy. The narrow scope of the Goldilocks zone is what makes us wonder if there is intelligent life other than our own. But, as previously stated, the number of terrestrial planets in our galaxy alone is enormous. Scientists now predict that 300 million planets in our Milky Way may have the ability to support life. The question is how many of these planets would have all the necessary ingredients for intelligent life to evolve? This number would likely drop substantially. The figure for intelligent life, similar to humans, could be nearer 500 to 1,000. The number of civilizations possessing advanced technology, including those who avoided extinction level events, may be far less. If there were just 25 advanced civilizations in our galaxy, we still must wonder why we have not seen evidence of their existence. Even if we are the only intelligent life in the Milky Way, there are still 2 trillion galaxies in our universe, making the odds for other intelligent life far beyond our comprehension.

There may be several reasons why we have yet to discover other intelligent life. We are all aware of the many possible threats to our continued presence on this planet. Most of these threats can be avoided, but there are some that could be considered extinction level events. This could be the reason we have yet to make contact with other intelligent life. These threats would include nuclear war, worldwide epidemics, asteroid impact, extreme solar activity, and extermination by artificial intelligence.

There are other factors that could delay contact with other life forms. With our current technology it would take over 6,000 years to travel to any of our closest stars. Alpha Centauri is a red dwarf star that is 4.37 light years away, which is 25.3 trillion miles from Earth. To make this trip in 6,000 years would require an average speed of 500,000 miles per hour. We are currently working on a future propulsion system called pulse fusion power, which would make this possible. However, 6,000 years would currently be impossible for human travel. Until we develop more advanced propulsion systems, only artificial intelligence could make these long journeys possible.

It has been less than 100 years since Edwin Hubble discovered galaxies beyond our own. It was not until 50 years ago that we could foresee that one day we might be able to travel beyond our solar system. As we become more technologically advanced, the threat to our existence becomes more apparent. We are just now beginning to address the threat of artificial intelligence. We can foresee how this form of intelligence will one day control every facet of our lives. This could be a real problem in the near future and the survival of the human race could be at stake. Just one miscalculation and a nuclear war could erupt without human provocation. All the challenges that the human race faces now and in the near future are the reasons why we may never discover other intelligent life in our galaxy. These same challenges will be common to any civilization no matter where they are located throughout the universe. What we do over the next two or three centuries may determine if the human race will survive. Once a civilization reaches a certain level of advancement, we can technologically develop safety nets to prevent some of the possibilities for extermination.

The search for alien life has been going on since the 1960s. Obviously, this is a short amount of time with respect to the existence of our planet. It was not even possible to receive information from a distant location in our galaxy until recently. But, considering the age of stars in our universe, you would think that there should be civilizations that are millions of years ahead of us. So why haven't we heard from ET? One possibility is that maybe there are no other alien life forms in the nearby portion of the galaxy we inhabit. I can't suggest that those areas beyond our reach do not have intelligent life. This is the second reason for not finding alien life. We do not have the equipment necessary to receive the signals we are searching for. Even if other intelligent life detected radio signals from Earth and they wanted to let us know they were there, they would need to be within 90 light years from Earth just to send us a message. We earthlings have only been transmitting radio signals for approximately 90 years. Even though signals travel at the speed of light, it still takes a lot of time to transit the vast distances between star systems. If we sent a message to let an alien race know we were here, however long it took them to receive it, it would take the same amount of time for them to respond. 25 years becomes 50, 50 years becomes 100. The other problem is that the other civilization would need to be looking in just the right place in our galaxy to receive our signal. Our galaxy is extremely large: 300 billion stars and 100,000 light years in diameter. Locating radio signals from Earth within this backdrop of immense space would be very difficult.

There are about 2,000 stars located within 50 light years from Earth. This is not a large number considering there are 300 billion stars in our galaxy alone. Out of these 2,000 stars, how many could harbor intelligent life? Very few, if not zero. Since intelligent life depends on so many near perfect conditions, the odds will require much larger numbers than 2000 stars. Most of these stars are red dwarfs, which likely do not have the ability to support life on any of their planets. Remember, the age of a star is important for life to evolve. So the number drops quickly when only stars similar to our Sun are necessary to generate intelligent life. Even if there were civilizations within this group of stars, they would need to possess our level of technology or better to be able to communicate in a way for us to know they existed. One of the greatest problems for not locating aliens is the fact that we have not been looking long enough. Unless we are looking in the right direction, at just the right time, and within a distance not far away, it will be only luck that will determine our success in locating other intelligent life. Another possibility to consider is that a more advanced intelligent life form may not care about looking for other life. They may have confirmed the existence of other alien life, but once they knew about its existence, they lost interest. They know that they cannot travel the great distances without a great number of years to make the voyage. They would have little interest in a culture that they believe to be in its intellectual infancy. If they were 1000 light years away, why make the trip if they had visited other civilizations in the past and experienced no advantage to make the long journey. In this case, this advanced alien life may decide to send robotic research spacecraft. Our only chance to discover this advanced alien race would be for us to receive signals produced by their society that were not intentionally transmitted.

A highly intelligent life form that is a million years more advanced than humanity would possess extraordinary mental skills. They would likely be biological and robotic. They would not be driven by emotion and would have little desire to communicate or make contact with our fragile and elementary society. Considering that our human race took 2 million years to reach the current technological stage, many alien life forms may be far more advanced. Two or three million years is nothing compared to the age of our galaxy. If it took 4.5 billion years for intelligent life to evolve, there could have been intelligent life within our galaxy several billions of years before humans. This is a far cry from our ancestry on Earth that may date back only 2 million years.

If all civilizations are as fragile as ours, the odds for survival are very small. However, an extremely advanced civilization will have learned how society should function and will have eliminated all the turmoil that creates unrest and division. Unless there was a purpose for improving their status quo, they may consider contact with another alien society to be a waste of effort. It is possible that we have been observed by alien life, but they chose not to make any contact based on our current intellect and cultural behavior. This may sound negative, but it is unfortunately relevant.

Most likely we will eventually find other alien life. This contact may not include a two-way conversation. It could be 200 years before we discover other life, but that may be the end of the story. I have often wondered how our society will react to this discovery. How will the discovery of ET play out? We have been searching for ET for over 60 years and so far, nothing. Now that we have AI assisting us in our search, maybe we will eventually be successful. We are now using very sophisticated algorithms to improve our search. These algorithms include technosignature elements, not just unusual radio signals. The signatures include what engineering has developed that is common to advanced technology. This includes laser pulses and high-tech radio signals. The technosignature also includes locating specific chemical concentrations in the atmosphere of a planet common to an advanced society.

Even with these advanced algorithms, it is extremely difficult to sort-out all the radio interference received by our radio telescopes. Trying to pinpoint the source is also difficult. Using an array of several radio telescopes may help in resolving this issue. Once we find a likely signal from ET, what do we do then? First of all, the entire scientific community would need to get involved to verify the accuracy of the findings. Then, all the world leaders would need to decide what to do next. Do we send a message to ET? Or, do we just keep listening to see if we can decipher more information about who or what this source may be? If we are smart, we will take our time and keep listening. It has been suggested by many in both science and government that we must be cautious in responding to any alien life form because they may be so advanced that there could be negative consequences. Stephen Hawking said we should be very careful about making contact with any more advanced alien life. Their intentions may be unpredictable.

If the source of ET was 300 light years away, they would not receive our message for 300 years. And, if they responded, it would be 300 more years for us to receive their message. If ET decided to visit Earth unannounced, it would likely be 3000 years before they arrived. By that time we might have the technology to deal appropriately with our newfound friend! Locating and meeting ET is not a simple feat. The distances in our universe are so vast that all forms of communication and travel are almost insurmountable.

If there is life in our galaxy other than the human race, it would likely evolve in a similar manner. Therefore, other intelligent life should not differ greatly from our own appearance. The reasoning for suggesting this similarity is based upon how life must endure the complexities of evolution. Specific requirements are necessary for survival of a species. There would always be some form of adversary that would challenge for control and superiority. This would include animal life and other intelligent life. Considering how humans evolved, it would be almost certain that other intelligent forms of life would evolve similarly in the same environment. To survive, intelligent life would require dexterity for making tools and weapons. They would require mobility to avoid danger and to gather food. The main differences in evolution would be natural selection and the gravity of the planet. The premise for how life evolves to an intelligent stage cannot vary greatly from what has transpired on Earth. Granted, life can exist in extraordinary circumstances, but for intelligent life to exist, the requirement for this evolution and survival are very narrow in scope. Based on the limited Goldilocks environmental requirements, there can be few civilizations throughout our galaxy. And because of these narrow limits, it is very likely that if another biological civilization exists, they will be physically very similar to the human race.

Even though extraterrestrials will likely be similar to humans, certain differences may exist that could be daunting. Due to different climates, natural selection and gravity of the planet, specific biological adaptations may exist. If aliens are from a smaller planet with less mass, they would be much thinner. Due to the lower gravity they would require less bone mass and muscle to perform any activity. They would not necessarily be taller. If they were from a planet slightly larger than Earth, they would likely be short and stocky.

Facial features, such as mouth, ears, and eyes should be very similar, however, aliens could have very different noses. The size of the nose is often related to climates. When humans evolve in a colder climate, the longer nose helps warm the air before it reaches the lungs. Our noses also increase our sense of smell. It is likely that ET will have a nose and not just slits for breathing, as is often portrayed.

They will also have similar eyes, not huge reptilian eyes which is not stereotypical of an intelligent life form.

The length of arms and legs of aliens could be slightly different, and it is possible they would not always be upright on just two legs. Depending on the variables of natural selection, they could walk and run on both their arms and legs. This may seem strange, but most likely they would be able to stand upright as well. Standing upright allows humans to use our arms and hands for improved flexibility and dexterity. They should also have fingers and toes for dexterity and stability. The length may vary depending on the gravity of their home planet. If their planet had a very mountainous terrain, longer appendages would improve their ability to climb.

The skin of an alien life form could also be very different because of climate and the luminosity of the home "star". Their skin could be thicker for better protection from possibly a harsher climate. It is surprising to me why humans evolved with such thin skin. Human skin is so vulnerable to variable weather, sharp objects, and radiation. Aliens may have more leathery skin for surviving a more radical climate and rough terrain of their home planet.

Humans have some hair on their body. Whether or not aliens will have hair on their body will be determined by their skin. The main reason humans have hair is to protect their skin. Hair can provide warmth, reduce chafing, and decrease exposure to radiation. Almost every depiction I have seen of ET reveals a naked hairless creature. ET would likely be from a planet with a variable climate, and I can assure you, they would not be running around naked all the time. If we meet ET someday, they will most likely be dressed with some type of loose clothing.

The discovery of alien life could be a challenge for many in the religious community. It has been almost 500 years since Copernicus published that the Sun, not the Earth, was the center of the universe. He had to wait until he was near death, in 1543, before publishing his heliocentric theory. Almost 100 years later, Galileo published the "Dialogue Concerning the Two Chief World Systems", which supported the heliocentric theory, obtained by observations he made with a new invention, the telescope. He was tried for heresy in 1632, and was forced to recant his findings to the Catholic Church. Today, the religious community has accepted many scientific findings, and it has not changed the message that the Church is providing. However, discovering other intelligent life should be comparable to what occurred 400 years ago. Discovering other alien life could be even more challenging, but we live in an era when the Church is more open to new discoveries in science. Still, how would this change the thinking of both the Church and society as a whole?

Today's Church has adapted and has accepted what science has proven to be fact. In the past, many in the religious community were also philosophers and scientists. As I discussed earlier, Georges Lemaitre was a Belgian Catholic priest who was the first to suggest the Big Bang theory. The Bible was created for the "good" of humanity. Science may include different perspectives concerning the purpose and function of the universe. And, as we know, the Bible and other religious writings were never meant to be a scientific study of the universe. The idea that the Earth was the center of the universe did not originate in the Church. The Earth center concept was originally suggested by Greek philosophers and was accepted by the Church as the preferred belief. Based on science around 500 BC, the Earth being the center of the universe made more sense than any other concept proposed during this time. The difference being, they had no telescope or other more advanced equipment to prove otherwise. It was just assumed that if God created "man", the Earth had to be the center of the universe. "Man" was a special creation by God and therefore he should dwell in a special place, the center of the universe.

Some of the earliest science dates back to ancient Egypt. Their astronomer priests studied the movements and positions of the stars. Around 2500 BC, these same priests devised the zodiac, breaking the sky into twelve groupings, each named after Egyptian gods or animals. Most Egyptians during this time believed in multiple gods. These gods controlled the sun, moon, rivers and almost everything else. They also had goddesses for wisdom, love, and plans for fighting battles. The Egyptians built temples to honor these gods, and the priests were in charge of managing the temples and events. They believed that several gods were responsible for creating the universe.

Around 150 AD, during the reign of Greek culture, philosopher Claudius Ptolemy created an elaborate chart which showed the Sun, Moon, and planets all orbiting around the Earth. This geocentric theory prevailed for 1400 years, until 1573 when Nicolaus Copernicus proposed the heliocentric theory with the Earth and other planets orbiting the Sun. This summary of how the Church and science were dependent on each other suggests that these two philosophies were not that far apart. It was not until the time of Galileo that the Church and science confronted each other's beliefs.

One of the biggest controversies between the Bible and science has always been how "man" was created. Even to this day, the Church, as a whole, does not support evolution as an explanation for our existence. The creationist's belief is that God created all living things, and they exist to this day the way they were created by God. Today's Catholic Church generally accepts the evolution of living things. This includes the concept of natural selection first proposed 1859 by Charles Darwin in his book "The Origin of Species". He was originally a creationist, but after many trips and extensive study of wildlife on the Galapagos Islands, he later realized how unique the animals and insects were in this isolated place. This is what changed his thinking about how life must have evolved differently on these remote islands. The Catholic Church and all other denominations would not accept this theory, considering it to be heresy. In 1950, Pope Pius XII of the Catholic Church was the first to reconsider its thinking on this topic, as well as the Big Bang theory. Most Catholic schools now teach evolution as part of their curriculum. Still most other religious affiliations do not support evolution to this day. Many of the younger generation have now become part of nondenominational churches and these churches are more open to the theory of evolution. The majority of people have become more aware of the vastness of the universe than ever before. At this point, locating other intelligent life is almost expected. It is just a matter of when.

In conclusion, life will go on as usual shortly after we discover there is other life in our galaxy. But, everyone on planet Earth will have a new sense of humility that will always be in the back of their minds. Many in the Church will still not accept evolution and suggest that God placed other alien life on their planet no different than was the case for human life. Others in the Church might possess a more open approach suggesting that evolution was all part of God's Plan. Since the Church normally will not alter a belief until it is proven to be fact, not theory, many may still refuse to accept the existence of other intelligent life unless there is actual contact. Direct contact with another alien civilization would likely affect everyone on the planet and completely change how we function as a society. We could only hope that the consequences would produce a positive outcome.

It is possible that for some reason humanity may be faced with circumstances that require we leave planet Earth in order to preserve our species. Elon Musk has said: "We must become a multi-planet species if humanity is going to survive". We are just beginning our venture into space. It will be difficult to overcome the many challenges of space travel, but we have to start somewhere. The biggest obstacle is the amount of space that is out there. The distances that would need to be navigated are currently beyond our reach. If we can develop propulsion systems and gravity simulation for extended space travel, distant exploration and eventually relocation could be possible. These space adventures would likely be one-way trips.

If we were to travel from planet Earth on these long ventures, we would experience a unique feeling of humility as our planet continued to become smaller and smaller until it was no different than all the stars in our field of view. As we pass Mars, Earth would appear to be a pale blue star, having a brightness similar to Jupiter (-2.7 magnitude), as seen from Earth. All that transpires on Earth will seem unimportant when we view this pale blue dot. As we continue our journey, the Earth will disappear from sight before we reach Neptune. Planet Earth will be only a memory. Without amplification, the Earth is lost in the darkness of space. Our Sun at this distance would appear like a very bright star. Someone with very sharp vision might be able to see the Sun as a small dot about the size of a large grain of salt held at arm's-length. The Sun's brightness on Neptune would appear like dim twilight. Continuing our journey beyond the outer reaches of our solar system, the Sun soon becomes just another star in the vastness of space.

As you leave our solar system, it's time to contemplate the significance of humankind. Our planetary system is in a minor spiral arm of our Milky Way galaxy lost among its 300 billion stars. If you were an alien traveler you would not likely find our solar system as you looked for planets that might contain intelligent life. The Milky Way is one galaxy among 2,000,000,000,000 other galaxies spread out over a distance of 92,000,000,000 light years or 540,000,000,000,000,000,000,000,000 miles. How difficult would it be to find our planet within this many miles of space, lost among 2 trillion galaxies? Why is the universe so large, as we begin to ponder the significance of the human race? We would all be just as happy if the universe was no larger than what it was believed to have been 2500 years ago. It would make no difference to us if it were that simple. Instead, we now know there may be no limit to how far the universe may extend, and we must consider there may be an infinite number of universes.

In order to address human significance in greater detail, I have to reference the title of this book. All that I have discussed in this book does not approach just how complicated the universe really is. When I reference the universe, I am also including the quantum world. The universe is so complicated, and it will never be fully understood. We have computers capable of solving almost any difficult equation, however, advanced technology is not capable of explaining infinity and multiple dimensions of space and time. Even if this were possible, we, as humans, would not be capable of deciphering the results. The scientific world is currently spending huge amounts of money in an effort to understand the universe, and it continues to become more complex.

This brings me back to the question: The Universe, why is it so complicated? We all take for granted our existence as humans. Only those who study the universe are aware of just how complicated it really is. In the fields of cosmology and physics it has actually become a game to search for just a "sliver" of proof for solving the riddle of the universe. Every time we make a new discovery, we think we made one step forward, but instead progress creates more challenges. It is like digging a hole to search for buried treasure, and as we continue to dig, the treasure just keeps falling deeper and out of reach.

Until we fully understand the universe, this question cannot be answered. But it can be stated that the more complicated the universe becomes, the greater the reason for our existence. **"We exist because the universe is so complicated!"** And, to answer the question about our significance is simple: We are here and part of a cosmic evolution that created our existence. We do not need to understand our purpose, but we have the opportunity to make a difference in humanity in the little time we have within a greater Plan. Maybe infinity will give us an opportunity to eventually understand our purpose as was orchestrated by our Creator.

Even though the universe is so complicated that we cannot understand it, the development of the human race is far more complex than any part of the universe. The human race is the ultimate creation for God. The fact that we have complex brains that can begin to comprehend the nature of the universe makes us special. Although I may have suggested that the universe may have cognitive abilities, there can be no comparison to the human mind. Our complexity is a marvel of creation. To repeat what Carl Sagan quoted at the beginning of this chapter: "We are an example of what hydrogen atoms can do, given 15 billion years of cosmic evolution".

Previously, I mentioned that another intelligent alien life form may not want to communicate with us. Thus, it would be difficult for us to discover other intelligent life in our galaxy. Unfortunately, our society has not advanced at the same level as our technology. If we, as a society, dedicated as much effort toward improving our society as we did our technology, we would become a civilization that might be more desirable for alien recognition. The next few centuries will likely be the turning point that will determine if we can survive as a human race.

So how can we change our society to avoid extinction? Our technology has progressed so rapidly over the past one hundred years. The human race has existed for such a brief time compared to the age of our universe. Three hundred thousand years ago we had just learned how to produce "fire". Today we are building spacecraft to explore our solar system. Where will humanity be three hundred thousand years from now? Most important, how will our society progress during this same period? There has been so little change in our society compared to five thousand years ago. What has made the situation worse is that there are now 8 billion people living on this planet. Avoiding conflict is far more difficult than ever before. Not only is conflict more likely, the amount of resources needed to care for 8 billion people is a major challenge. This picture of our circumstance for human survival is the reason why we, as a society, must make a greater effort to develop a plan to improve how we work together as a people. If we don't change, our survival is in question. If humanity could learn to work together toward the goal of a better life for everyone, there would be less cause for dispute. If we could learn better methods for treating mental health, we could prevent discontent and aggressive behavior. If we could learn to understand and respect different cultures, we could work together as nations. If children were taught morality in our schools, most conflicts could be avoided throughout life. If nations could respect each other, they would prosper far more by sharing their ideas and skills for creating more resources. If our United Nations would become just that, war would no longer be considered! If we can accomplish most of these goals, humanity will survive the next three hundred thousand years and beyond.

During this time, many technological advances will drastically change the framework of the human race. We are now in the early stages of robotics. Over the next two hundred years, this will change how we exist in our society in so many ways. Not only will our manufacturing be done entirely with robotics, even our household chores will no longer require our personal attention. Driving a car will become a thing of the past. With robotics dominating so much of our lives, we will ultimately become more robotic in our own way. In the not too distant future, specific computerized implants may be inserted into our brains to help us function better in our daily lives. This could become a turning point in our society as to whether or not this is accepted as a normal process. In the future we will have to be careful how we choose to incorporate technology into our lives. This may be the biggest challenge for the preservation of humanity.

Robotics could be used for many beneficial reasons. Mental illness could be corrected with a chip inserted into the brain to regulate thoughts and emotions. We are already incorporating bionic arms and legs for those who have lost limbs. Transplanting artificial organs may also be common. By using these technological advancements we will also live much longer lives. Life spans may exceed 200 to 300 years in the near future, making space travel far more appealing for those who choose to pursue this endeavor. What concerns me most about incorporating bionics is how this technology could be used in our brains. We cannot lose our ability to make choices, feel emotions, and to experience a sense of gratification for attaining goals with hard work. We must separate ourselves from robots or we may lose the main components for human existence.

The greatest challenge for meeting another alien race will be the difference in technological advancement. It is impossible for us to imagine the human race one million years in the future. How will our society be different in the year 3000? How will our society be different in the year 1,002,024? This is beyond our ability to predict! This could well be the circumstances for our confronting an alien race.

Considering how advanced an alien race would need to be to travel thousands of light years, it would likely require that the alien race be some form of artificial intelligence.

If we can preserve humanity for many thousands of years, we may desire to make the journey to parts of our own galaxy. Even with extraordinary technology, these journeys could be very difficult. Remember, our galaxy is over 100,000 light years in diameter. Just to travel 5,000 light years at 10% the speed of light (67,000,000 mph) would take approximately 50,000 years. Unless some type of "wormhole" travel is available in the future, these distances may never be considered by any biological species. Only robots could make distant travel possible. These same robots could transport the human species in some form, just to seed a new planet with human life. This may be the only way to colonize our own galaxy beyond a few hundred light years.

On a grander scale, the distance between galaxies is beyond the reach of almost any civilization, no matter how advanced the technology. These journeys would take tens of millions of years, and no spacecraft could survive the journey. Considering the time needed to journey to any part of our own galaxy, it truly becomes more apparent why we may never meet another biological intelligent life form. For space travel, the size of our universe is far beyond anything we could imagine.

As I suggested, the only way to colonize another planet over 300 light years from Earth would be for AI to transport human life in some preservative form (possibly frozen embryos). Traveling thousands of years in space would be nearly impossible for biological intelligent life. The question is whether this would even be considered by a very advanced civilization? If AI were the controlling force on this or any other planet, would this be a priority? Most likely not.

There has been some speculation that the human race was placed on Earth in the distant past. There are reasons that would discount this possibility. First of all, there have been no remnants found that were left behind by a visiting alien race. Even if AI aliens planted the seed for the human race, they would have stayed for a while to monitor the development of this new form of humanity. If this action occurred it would likely have taken place approximately 6 to 7 million years ago. According to recent findings, this is when the human race genetically branched off from similar primates. This is the only time in the evolution of the human race that is not fully understood. The evolution of humans on Earth has otherwise been proven to follow a specific chain of events that cannot be disproven. Humans share DNA comparisons that are not that different from other primates, which would make it more likely that humans evolved here on Earth. If an alien race did "plant the seed" for humanity, they would come back to monitor the progress. Although there have been "sightings" of UFOs, there has been no definite confirmation of aliens visiting Earth.

The second problem is the sheer distance that would need to be navigated to carry out this mission. Even AI must have a reason to travel 50,000 years or more to an unknown planet, Earth. Why attempt such a challenging journey? Unless the journey was less than 500 light years, I cannot fathom their consideration for this type of endeavor. Even a distance of 500 light years would take 5000 years at 10% the speed of light. There would be many challenges for a spacecraft to travel these distances for thousands of years.

Reality is sometimes hard to accept. Our existence may be rare, and we may never meet other biological intelligent life. This may have been God's Plan, because the place we live, the Universe, is beyond the scope of humanity.

To colonize this planet near Orion, 1500 light years from Earth, would require 15,000 years just to make the journey, at 1/10 the speed of light. Artwork by Larry Eichenauer (Clay/ Projection)

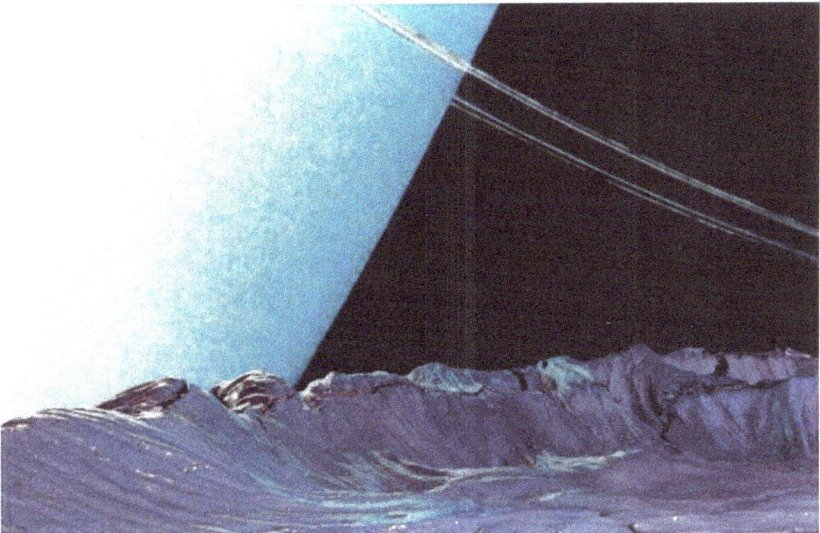

Very few exoplanets may possess the near perfect conditions required for intelligent life to evolve. Life similar to humans may be extremely rare. Artwork by Larry Eichenauer (Clay/ Projection)

Traveling to other galaxies is far beyond the reach of any biological or artificial form of life.

Artwork by Larry Eichenauer (Clay/ Projection)

Life may exist in other parts of our galaxy, but making the exploratory journey may be too difficult for humans.

Artwork by Larry Eichenauer (Clay/ Projection) *Projections: NASA

Final Thoughts

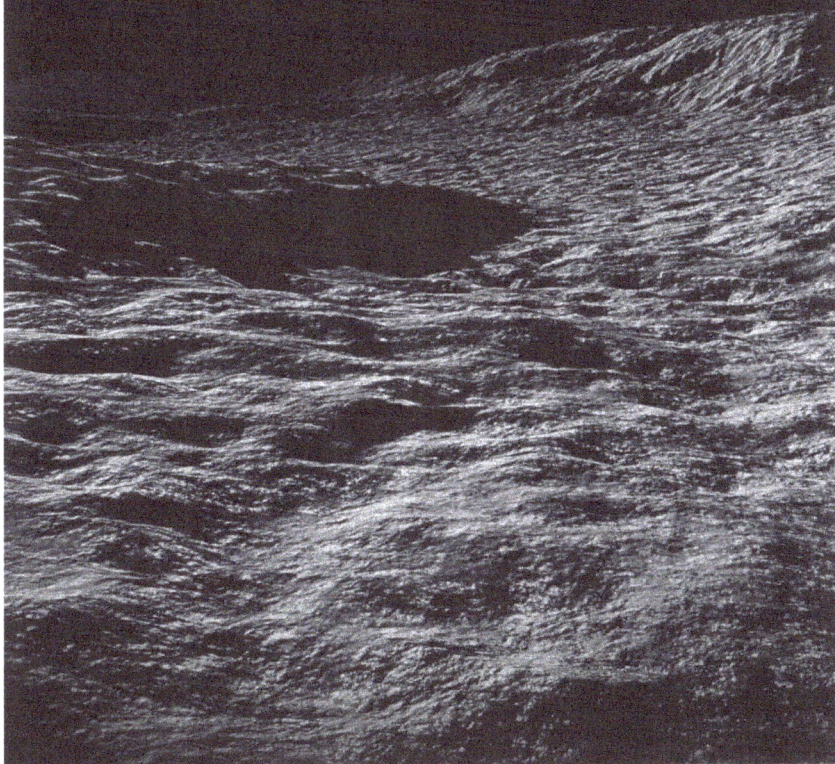

In the chapter, "Black Hole Endoverse Theory", I discussed briefly the possibility of infinite universes. When one considers the complexity of our own universe, it is difficult to imagine an endless number of universes all connected by a common thread. Our universe is so vast, and yet, when this enormity is repeated forever for all eternity, it becomes incomprehensible. Why does all THIS exist? How could something this complicated come into being? Why would the Creator of the Universe design something beyond the scope of imagination? Each of us is a part of this infinite Plan, but we live less than a century to witness this prodigious creation. What lies beyond our life on Earth is an even greater mystery. God has a purpose for everything, but we will never understand how humanity is part of this infinite structure. The biggest question of all is, WHY did God create this inconceivable complex infinite number of universes? The WHY is as baffling as the HOW.

There are many that believe that all THIS was created just for the existence of humanity. This premise may be no different than the earth centric belief two thousand years ago. However, if there are millions or even billions of intelligent life forms throughout the Universe, this could be one answer for WHY. If just one human-like civilization existed in every galaxy, there could be two trillion intelligent life forms in our universe. If this is then part of an infinite number of universes, intelligent life would be countless. Since there is such a great distance between galaxies, each alien race would be able to colonize their own galaxy and possibly <u>never</u> encounter any other type of intelligent alien life. Could God have "staged" one massive black hole at the center of every galaxy for the purpose of re-creating the universe, and also provide the means for at least one intelligent life to evolve within each galaxy? This could be the Master Plan that would answer the question for, WHY?

Many great minds believe there must exist a Force or Power that is much greater than humanity. Einstein did not deny the existence of God. There have been few statements recorded based on Einstein's thinking about God. He chose to refrain from comments concerning this topic. He did say he was not an atheist. Nikola Tesla, a well-known physicist and engineer, said: "What one man calls God, another calls the laws of physics". Even though this viewpoint is shared by some scientists, over fifty percent in the field of science believe in God as a superior force of energy, and the primordial Creator of the Universe.

As a whole, most people believe in God. The following reveals the percentage of humanity that believe in God, plus those who are uncertain or do not believe: Christianity 35%, Islam 25%, and Hindu 15%. For the remaining, 10% are agnostic, and 15% do not believe in God (8% Buddhist, 7% atheist). Buddhists believe in an endless cycle of reincarnation. They believe the universe continues a cycle of expansion and contraction. For some, these numbers may be disappointing, but, 75% of the human race believing in God is very significant. Since the beginning of civilization, humanity has had the desire to believe in something greater than themselves. Most of humanity chooses to believe we have a purpose in this vast universe. For many, a lack of understanding encourages our belief in a Higher Power. At the same time, the more we learn, it is apparent there is something so much greater than "mankind" that produced such an elaborate Creation.

One of the most significant topics among all those who believe in God is "life after death". For those who share this belief, death is just another aspect of life. Their beliefs and actions here on Earth will determine their life in Heaven. So, what do we know about life after death? Obviously, we know nothing about what happens after we die. Those who have been dead for minutes, and then survived, have similar stories. Most have seen themselves hovering above their own body immediately after death. Some have seen a very bright light and sensed an extreme calmness. Some have seen family members surrounded by a bright light. All these visions occur within minutes after death, but what happens beyond this small period of time is a total mystery. From a religious standpoint, one just needs to have faith in the belief of life after death, as advocated in the Bible, and other theological teachings. In Ecclesiastes 12:7, the Bible states about death, "Then shall the dust return to the earth as it was; and the spirit shall return to God who made it".

From a scientific viewpoint, is eternal life or life after death possible? In other words, would God provide a method for this to occur? First of all, we do not know what we become in "spirit". This is part of the mystery of the unknown. Our spirit (soul) may be an energy force that is released upon our death. What happens to this force? "God only knows!" Scientifically, our spirit might enter a different dimension of space and time. God could have provided a specific dimension just for the spirits of those who die. One of the most significant questions for all humanity: Did God create places called Heaven and Hell? Those who study the Bible will maintain that if people follow God's word, they will be rewarded with a place in Heaven.

If our existence is unique and special, there would be reason to believe God would have provided a means for us to enjoy this vast creation more than a single lifetime. We know nothing about other dimensions of space and time, so the possibilities could be endless. If our Creator had the power to "build" a Universe, I cannot imagine what else could be possible.

Just how important are we in God's Creation? Are we that significant to such a degree our Creator would give us greater favor over the rest of this vast creation? Considering we were born on a remote planet, lost among countless stars within two trillion galaxies and an infinite number of universes, is it conceivable we would be the most important part of the Plan of our Creator? It might seem difficult to imagine this, but, it is possible.

Throughout this book I have suggested just how complicated the Universe must be for the existence of the human race. The human evolutionary process is not possible without a pattern of events that must occur with near perfect conditions. The essential ingredients for our existence are far more specific than what we could imagine. Some of these conditions include a star (our Sun) that must have moderate fluctuations in luminosity and radiation. Our planet is located at the correct distance from the Sun for life to evolve. The Earth's composition of water and its metal core are essential for life. Earth's metal liquid core produces a precise magnetic field that provides protection from the Sun's radiation. Our Moon is responsible for helping maintain stability of the Earth's axis, and its tidal effects control the movement of water currents over our planet's surface. Without these tidal influences, the evolution of life would have been greatly altered. Without the Moon, intelligent life may not have evolved on our planet. If the Earth was not tilted on its axis, our existence and survival may have not been possible. Without seasons, food supply would be limited, and the spread of disease would be more prevalent from additional insects and bacteria. And finally, for DNA to evolve for the development of human life, so many events must occur in a specific sequence to create the precise evolutionary process.

All these extraordinary requirements for existence must suggest that humanity is not just a coincidence. We exist as a result of a near perfectly orchestrated Plan. Every last detail may be essential for almost any intelligent life form to evolve in our Universe. We cannot take for granted all that is necessary for our existence. Humanity or any intelligent life form will not exist in our Universe without our complicated quantum world, exploding stars, black holes, and stars like our Sun. Even though the Universe is so complicated and immense, we must not let this overshadow the reason for our existence. God must have a purpose for the human race.

We must accept the obvious. The universe exists for reasons we will never know. But one thing is certain: We are here to witness and be a part of, what GOD has created.

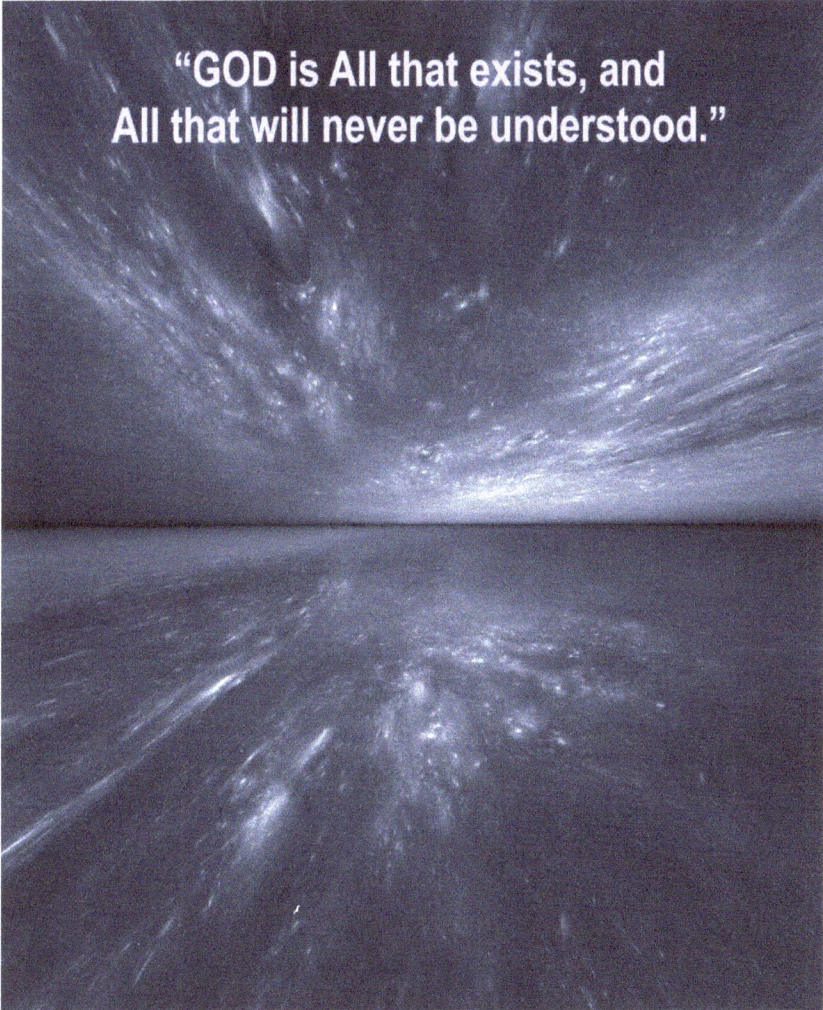

"INFINITE HORIZON"
Quote: Larry Eichenauer

Acknowledgements

Reference Literature that provided some quotes and thoughts in the text of this book:

"The Universe" by John Brockman

"You Are The Universe" by Deepak Chopra, MD

"Wikipedia" for information of relevant facts and statistics

Quotations at the beginning of each chapter were obtained online from 3 sources: Brainyquote.com, Goodreads.com, and wwu.edu>cosmo

All Photos: iStockPhoto.com

Front Cover – den-belitsky

Back Cover – angel_nt

Ch. 1 – In the Beginning – pixelparticle

      Big Bang 1 – ttsz, **Big Bang 2** – suirey

Ch. 2 – Universe Head – WhataWin, **Cosmic Web -** Shooter99, **Black Hole** – Elen11,**Quasar** – Elen11

Ch. 3 – Black Hole - Daniel Megias, **Hole -** gremlin & fotojog

      Wormhole – gremlin **& Universe** – fotojog

Ch. 4 – Dark Matter – AntonioSolano, **Neurons** – whitehoune

      Cosmic Web – Shooter99

Ch. 5 – Atom – rendixalextian, **Molecule to Quark** – ttsz

      Gravity 1 – sakkmesterke, **Gravity 2** – graphics.vp,

      Gravity 3 – somboon sitthichoptam

Ch. 6 – Earthlite – fpm

Final Thoughts – Earth/moon - magann

Infinite Horizon - ppart

About the Author

Larry Eichenauer graduated from Ohio Northern University, and continued postgraduate studies at Prairie View A&M University, including additional courses in Astronomy at the University of Houston. His lifetime interest and extensive research in the fields of physics and cosmology combined with a creative philosophy has generated some of the most original theories for the creation and function of the universe. Some of these groundbreaking ideas could inspire a new direction for future research. These novel theories may provide answers for explaining dark matter and dark energy.

In addition to physics and cosmology, the author has a technical and creative background. He holds 4 patents and has numerous copyrights and trademarks. His writing career is fairly recent. His previous books are written about golf, advocating the mind, rather than the body, determines how well an individual performs. The author also has a background in art and design. Some of his artwork is displayed in this book.